Study Guide Workbook

Glencoe

Geometry

Concepts and Applications

New York, New York Columbus, Ohio Woodland Hills, California Peoria, Illinois

To the Teacher:

Answers to each worksheet are found in Glencoe's *Geometry: Concepts and Applications Study Guide Masters* and also in the Teacher's Wraparound Edition of Glencoe's *Geometry: Concepts and Applications.*

Glencoe/McGraw-Hill

*A Division of The **McGraw-Hill** Companies*

Send all inquiries to:
The McGraw-Hill Companies
8787 Orion Place
Columbus, OH 43240-4027

ISBN: 0-02-834826-5

Study Guide Workbook

5 6 7 8 9 10 024 07 06 05

Contents

NAME ______________________________ DATE __________ PERIOD ______

1-1

Study Guide

Student Edition
Pages 4–9

Patterns and Inductive Reasoning

In daily life, you frequently look at several specific situations and reach a general conclusion based on these cases. For example, you might receive excellent service in a restaurant several times and conclude that the service will be good each time you return.

This type of reasoning, in which you look for a pattern and then make an educated guess based on the pattern, is called **inductive reasoning**. The educated guess based on these facts is called a **conjecture**. Not all conjectures are true. When you find an example that shows that the conjecture is false, this example is called a **counterexample**.

Example: Find the next three terms of the sequence 3, 8, 13, ...

Study the pattern in the sequence.

3, 8, 13, . . .

+ 5 + 5

Each term is more than the term before it. Assume that this pattern continues. Then find the next three terms using the pattern of adding 5.

3, 8, 13, 18, 23, 28

+ 5 + 5 + 5 + 5 + 5

Find the next three terms of each sequence.

1. 17, 25, 33, . . .

2. 60, 52, 44, . . .

3. 2, 6, 18, . . .

4. 7, 9, 13, . . .

5. 11, 7, 3, . . .

6. 24, 12, 6, . . .

7. Find a counterexample for the statement.
All animals have fur.

NAME ______________________ DATE __________ PERIOD ______

Study Guide

Student Edition
Pages 12–17

Points, Lines, and Planes

Term	Description	Names
point	• has no size	point P
line	• is an infinite number of points	line m or line AB or $\overleftrightarrow{AB}$
ray	• starts with a point called an **endpoint**	ray CD or $\overrightarrow{CD}$
line segment	• is part of a line with two endpoints	line segment EF or $\overline{EF}$
plane	• is a flat surface that extends without end; has no depth.	plane $\mathcal{G}$ or plane GHI

Use the figure at the right to name examples of each term.

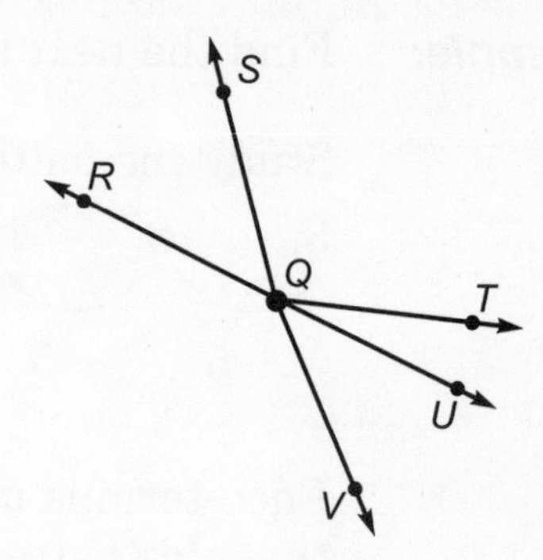

1. ray

2. point

3. line

4. line segment

The diagram at the right represents a baseball field. Name the segment or ray described in Exercises 5–9.

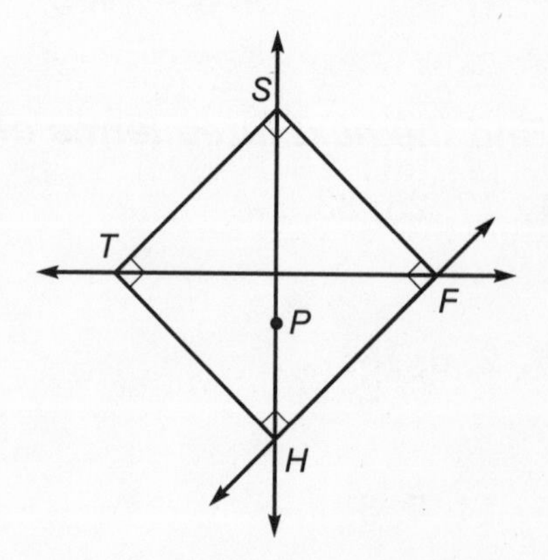

5. The player on third base throws the ball and the player on first base catches it.

6. The batter hits the ball over the head of the player on first base.

7. The player on first base throws the ball home, but the catcher misses it.

8. The batter hits the ball and it is caught at second base.

9. Are third base, first base, and the pitcher's mound collinear?

Study Guide

Postulates

Point, line, and plane are undefined terms in geometry. The **postulates** describe the fundamental properties of these terms.

A club is divided into committees. The undefined terms are *committee* and *member*.
Postulate 1: Each pair of committees has exactly one member in common.
Postulate 2: Each member is on exactly two committees.
Postulate 3: There are exactly four committees.

Use the postulates above to complete Exercises 1–3.

1. Draw a model to illustrate these rules. Use circular shapes to represent committees and letters to represent members.

2. How many club members are there?

3. How many members are on each committee?

Suppose some companies have collaborated to place several satellites in orbit. Let the set of all satellites that a given company helped place in orbit be called a *network*.
Postulate A: There are at least two distinct satellites.
Postulate B: For each pair of satellites, there is exactly one network containing them.
Postulate C: Each network contains at least two distinct satellites.
Postulate D: For each network, there is a satellite not in it.

Use the postulates above to complete Exercises 4–6.

4. Draw a model to illustrate these rules. Use rectangles to represent networks and letters to represent satellites.

5. What is the least number of satellites?

6. What is the least number of networks?

NAME ______________________ DATE ________ PERIOD _____

Study Guide

Student Edition
Pages 24–28

Conditional Statements and Their Converses

If-then statements are commonly used in everyday life. For example, an advertisement might say, "If you buy our product, then you will be happy." Notice that an if-then statement has two parts, a *hypothesis* (the part following "if") and a *conclusion* (the part following "then").

New statements can be formed from the original statement.

Statement	$p \rightarrow q$
Converse	$q \rightarrow p$

Example: Rewrite the following statement in if-then form. Then write the converse, inverse, and contrapositive.

All elephants are mammals.

If-then form: If an animal is an elephant, then it is a mammal.
Converse: If an animal is a mammal, then it is an elephant.

Identify the hypothesis and conclusion of each conditional statement.

1. If today is Monday, then tomorrow is Tuesday.

2. If a truck weighs 2 tons, then it weighs 4000 pounds.

Write each conditional statement in if-then form.

3. All chimpanzees love bananas.

4. Collinear points lie on the same line.

Write the converse, of each conditional.

5. If an animal is a fish, then it can swim.

6. All right angles are congruent.

NAME ______________________ DATE __________ PERIOD _____

Study Guide

Tools of the Trade

A **straightedge** is any object that can be used to draw a straight line. A **compass** is any object that can be used to draw circles.

The figures in Exercises 4–7 below were constructed using a straightedge. A figure can be traced if it has no more than two points where an odd number of segments meet.

Use a straightedge or compass to determine whether each statement is true or false.

1. Circle P is larger than circle Q.

2. Circle P is smaller than circle Q.

3. The two circles are the same size.

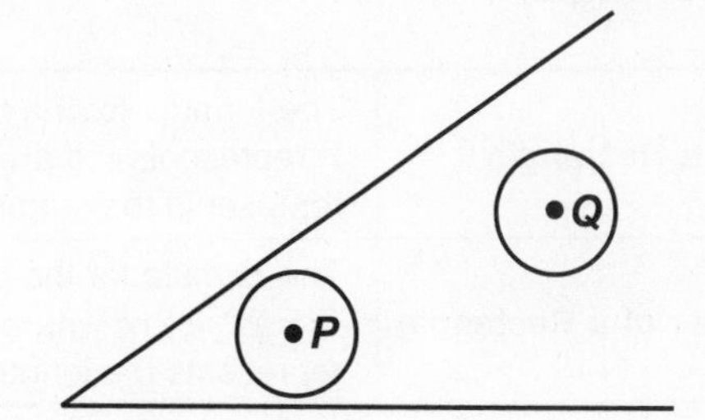

Determine whether each figure can be traced. If so, name the starting point and number the sides in the order they should be traced.

4.

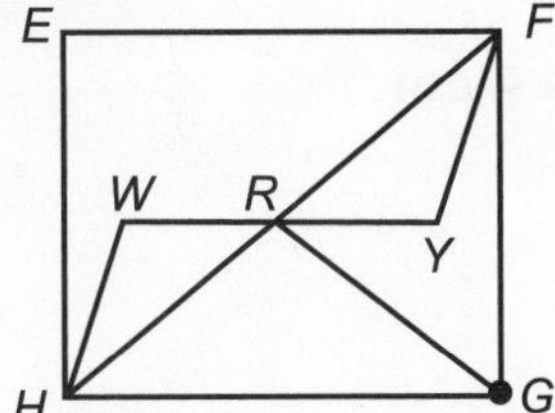

5.

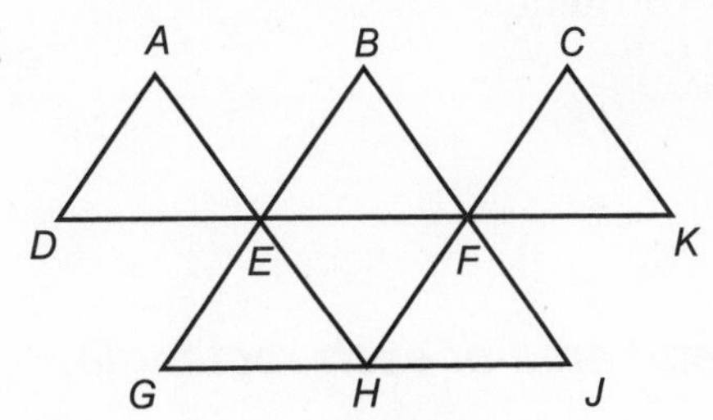

6.

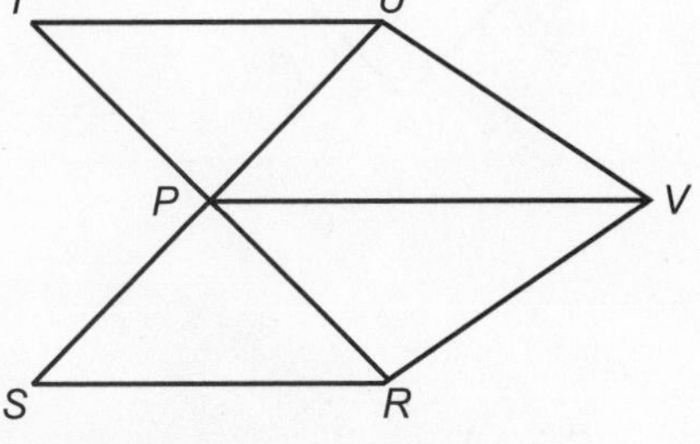

7.

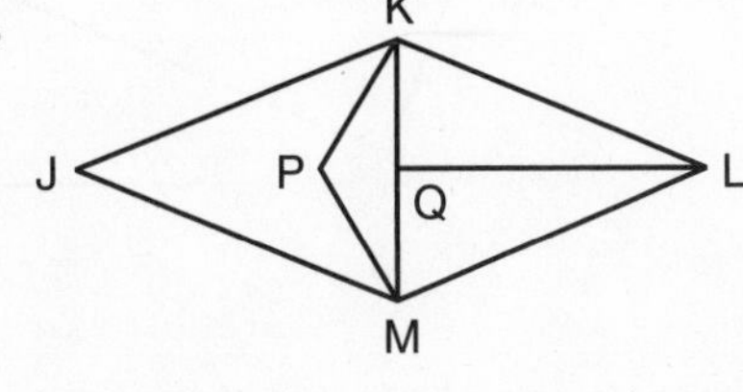

NAME ______________________ DATE ________ PERIOD _____

Study Guide

A Plan for Problem Solving

The following four-step plan can be used to solve any problem.

Problem-Solving Plan	
1. *Explore* the problem.	Identify what you want to know.
2. *Plan* the solution.	Choose a strategy.
3. *Solve* the problem.	Use the strategy to solve the problem.
4. *Examine* the solution.	Check your answer.

When finding a solution, it may be necessary to use a formula. Two useful formulas are the area formula and perimeter formula for a rectangle.

Area of a Rectangle	The formula for the area of a rectangle is $A = \ell w$, where A represents the area expressed in square units, ℓ represents the length, and w represents the width.
Perimeter of a Rectangle	The formula for the perimeter of a rectangle is $P = 2\ell + 2w$, where P represents the perimeter, ℓ represents the length and w represents the width.

Examples

1 Find the perimeter and area of the rectangle at the right.

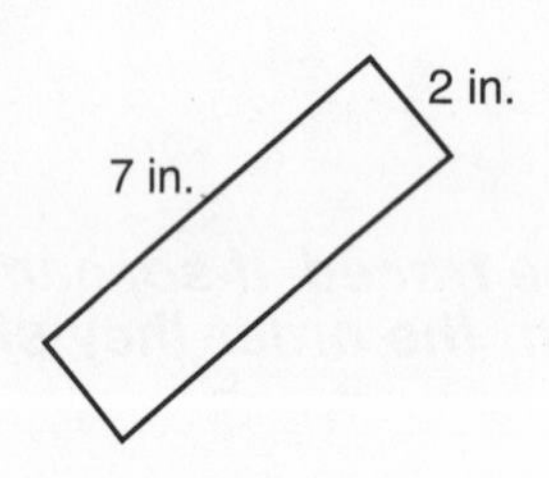

$P = 2\ell + 2w$
$= 2(7) + 2(2)$
$= 14 + 4$ or 18
The perimeter is 18 inches.

$A = \ell w$
$= 7 \cdot 2$ or 14
The area is 14 in^2.

2 Find the width of a rectangle whose area is 52 cm^2 and whose length is 13 cm.

$A = \ell w$
$\frac{52}{13} = \frac{13w}{13}$
$4 = w$
The width is 4 cm.

Find the perimeter and area of each rectangle.

1.

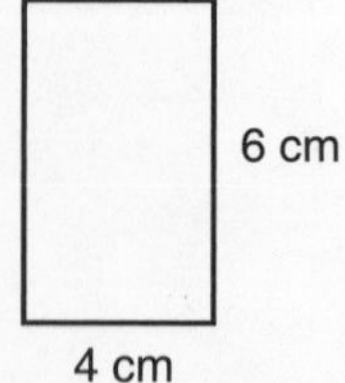

2.

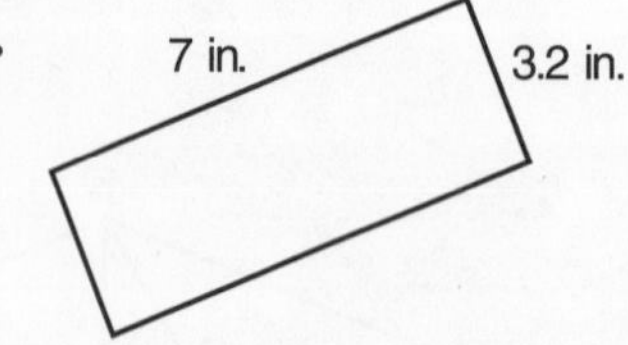

3.

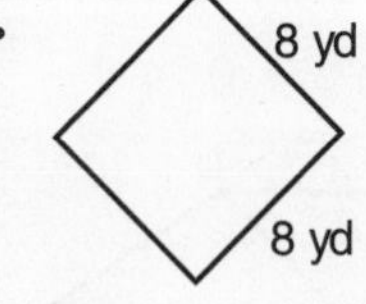

Find the missing measure in each formula.

4. $\ell = 3, w = 7, P = \underline{?}$

5. $w = 5.2, \ell = 6.5, A = \underline{?}$

6. $w = 4, A = 36, \ell = \underline{?}$

7. $P = 65, \ell = 18, w = \underline{?}$

NAME ______________________ DATE __________ PERIOD ______

2-1 Study Guide

Student Edition
Pages 50–55

Real Numbers and Number Lines

Numbers can be grouped into sets with identifying characteristics.

Sets of Numbers		
Name	**Definition**	**Examples**
whole numbers	0 and the natural, or counting numbers	0, 1, 2, 3, . . .
integers	0, the positive integers, and the negative integers	. . . −3, −2, −1, 0, 1, 2, 3, . . .
rational numbers	any number of the form $\frac{a}{b}$, where a and b are integers and $b \neq 0$	$\frac{1}{3}$, 7.9, $2\frac{5}{8}$, 9.3686868 . . .
irrational numbers	decimals that neither terminate nor repeat	0.513947836 . . . , 1.010010001 . . .
real numbers	rational and irrational numbers	$4.\overline{68}$, $\frac{9}{11}$, −21.494994999 . . .

Each real number corresponds to exactly one point on a number line. The distance between two points on a number line is the positive difference of their coordinates.

For each situation, write a real number with ten digits to the right of the decimal point.

1. a rational number between 5 and 6 that terminates

2. an irrational number between 1 and 2

3. a rational number between –3 and –2 with a 3-digit repeating pattern

Use the number line to find each measure.

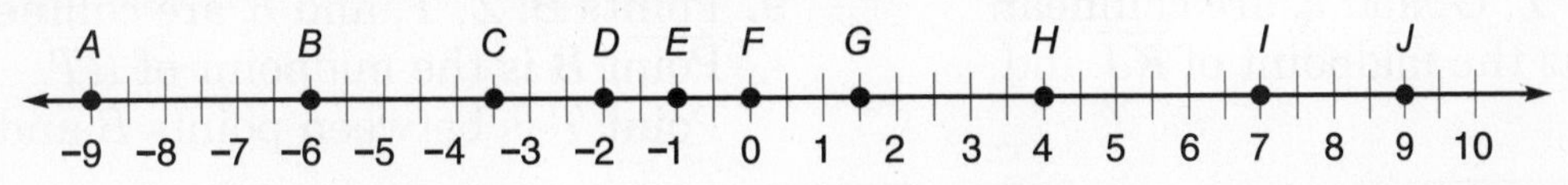

4. *HI* **5.** *AD* **6.** *BH* **7.** *AJ*

8. *BC* **9.** *CG* **10.** *CJ* **11.** *FC*

NAME ______________________ DATE __________ PERIOD ______

Study Guide

Segments and Properties of Real Numbers

Point B is **between** points A and C if A, B, and C are collinear and $AB + BC = AC$.

Three segment measures are given. The three points named are collinear. Determine which point is between the other two.

1. $NQ = 17, NK = 6, QK = 11$

2. $JB = 9.8, BP = 3.2, JP = 6.6$

For each of the following, draw the missing point. Then use the information given to write the coordinates of all three points.

3. $CD + DF = CF, CD = 3, DF = 5$
   The coordinate of C is 20.
   C D

4. $MK + KP = MP, MK = 6, KP = 4$
   The coordinate of M is –25.
   P K

5. $QB + BA = QA, AB = 10, QB = 8$
   The coordinate of B is 1.
   A B

Draw a figure for each situation described.

6. Points M, A, N, and Q are collinear. Point N is between points A and Q. Point M is not between points A and Q.

7. Points N, Q, W, C, and E are collinear. Point N bisects $\overline{CE}$. Point W bisects $\overline{NE}$. Point Q is the midpoint of $\overline{WE}$.

8. Points J, Y, G, and K are collinear. Point G is the midpoint of $\overline{KJ}$ and $\overline{KJ} \cong \overline{JY}$.

9. Points B, Z, T, and R are collinear. Point B is the midpoint of $\overline{RT}$. Point T is between points B and Z.

NAME ______________________ DATE __________ PERIOD ______

Study Guide

Student Edition
Pages 62–67

Congruent Segments

Two segments are congruent if they have the same length. The **midpoint** of a segment separates the segment into two congruent segments. To **bisect** a segment means to separate it into two congruent parts. The midpoint always bisects a segment.

Use the line to name all segments congruent to each given segment.

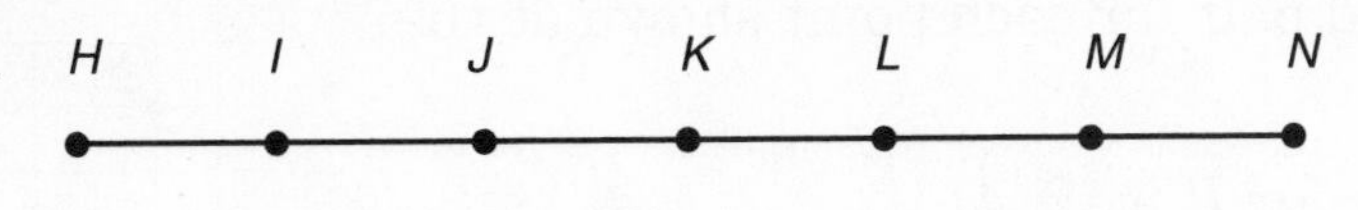

1. $\overline{HM}$

2. $\overline{JL}$

3. $\overline{NJ}$

4. $\overline{HI}$

Use the number line to name the midpoint of each segment.

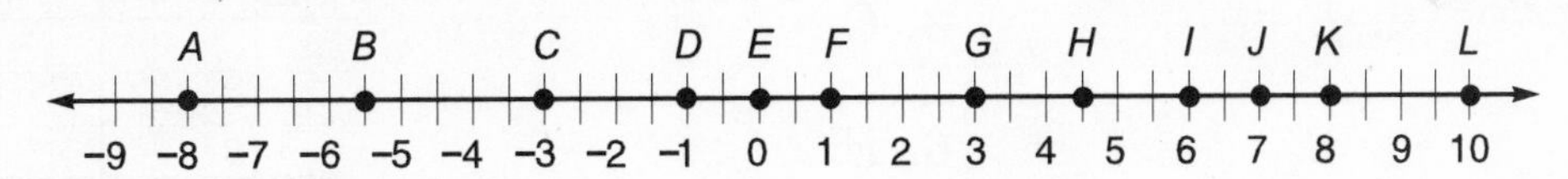

5. $\overline{EI}$

6. $\overline{IL}$

7. $\overline{AK}$

8. $\overline{CF}$

9. $\overline{AC}$

10. $\overline{DL}$

11. $\overline{CG}$

12. $\overline{IK}$

For each exercise below, the coordinates of points P and Q, respectively, are given. Graph P and Q. Then draw and label the coordinate of the midpoint of $\overline{PQ}$.

13. –4 and 2

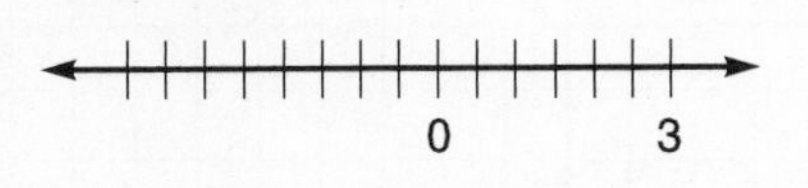

14. –9 and –5

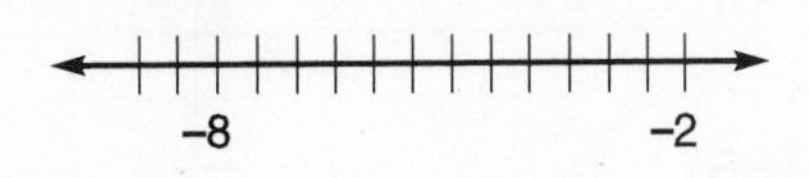

15. –3 and 4

NAME ______________________________ DATE __________ PERIOD ______

2-4 Study Guide

Student Edition
Pages 68–73

The Coordinate Plane

Every point in the coordinate plane can be denoted by an ordered pair consisting of two numbers. The first number is the ***x*-coordinate**, and the second number is the ***y*-coordinate**.

> To determine the coordinates for a point, follow these steps.
> 1. Start at the origin and count the number of units to the right or left of the origin. The *positive direction* is to the right, and the *negative direction* is to the left.
> 2. Then count the number of units up or down. The positive direction is up, and the negative direction is down.
>
>   *Note:* If you do not move either right or left, the *x*-coordinate is 0. If you do not move up or down, the *y*-coordinate is 0.

Example: Write the ordered pair for each point shown at the right.

The ordered pair for R is (2, 4).
The ordered pair for S is (−3, 3).
The ordered pair for T is (−4, −2).
The ordered pair for U is (1, −4).
The ordered pair for W is (0, 2).
The ordered pair for X is (−2, 0).

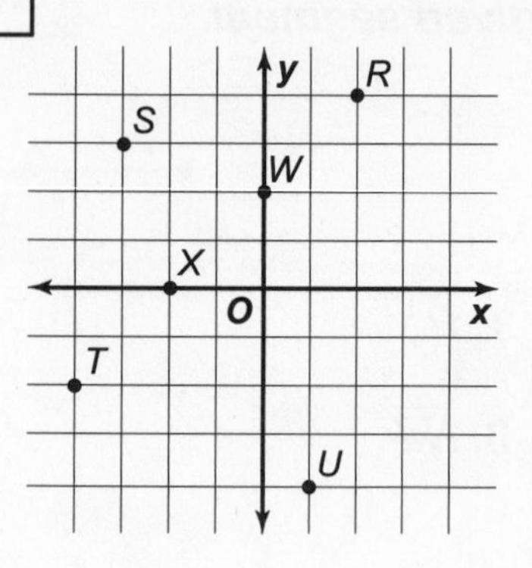

Write the ordered pair for each point shown at the right.

1. A **2.** B **3.** C

4. D **5.** E **6.** F

7. G **8.** H **9.** I

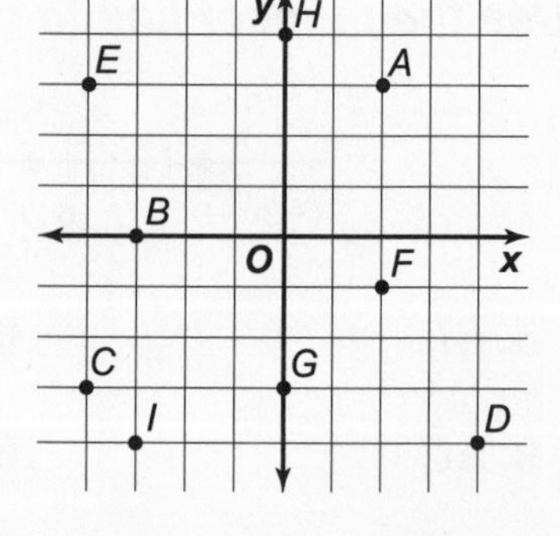

Graph each point on the coordinate plane.

10. $M(6, 4)$ **11.** $N(-5, 4)$

12. $P(-3, 5)$ **13.** $Q(6, 0)$

14. $J(0, -4)$ **15.** $K(7, -5)$

16. $Y(9, -3)$ **17.** $Z(-8, -5)$

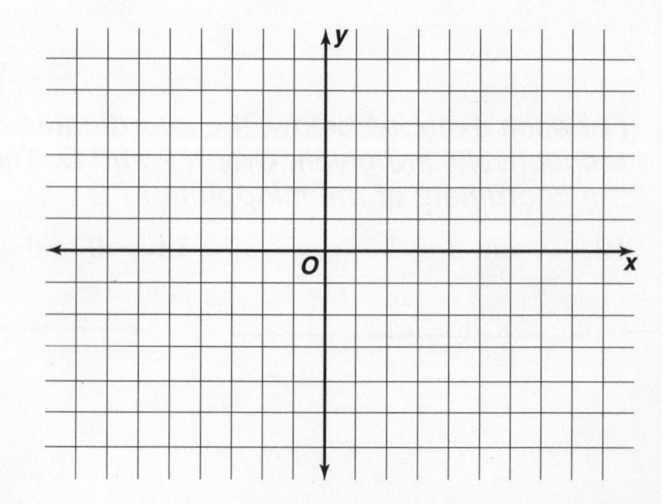

NAME ______________________ DATE ________ PERIOD ______

2-5 Study Guide

Student Edition
Pages 76–81

Midpoints

There are two situations in which you may need to find the midpoint of a segment.

Midpoint on a Number Line	Midpoint in the Coordinate Plane
The coordinate of the midpoint of a segment whose endpoints have coordinates a and b is $\frac{a+b}{2}$.	The coordinates of the midpoint of a segment whose endpoints have coordinates (x_1, y_1) and (x_2, y_2) are $\left(\frac{x_1+x_2}{2}, \frac{y_1+y_2}{2}\right)$.
Example 1:  The coordinate of the midpoint of $\overline{RS}$ is $\frac{-3+9}{2}$ or 3.	***Example 2:*** 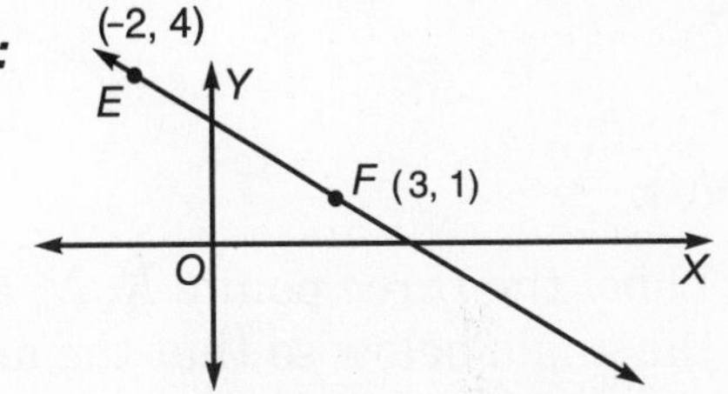 The coordinates of the midpoint of $\overline{EF}$ are $\left(\frac{-2+3}{2}, \frac{4+1}{2}\right)$ or $\left(\frac{1}{2}, \frac{5}{2}\right)$.

Use the number line to find the coordinate of the midpoint of each segment.

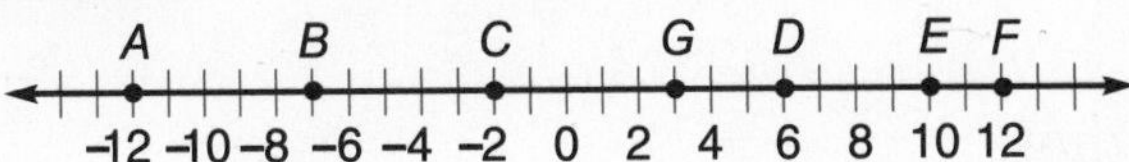

1. $\overline{AB}$ **2.** $\overline{BC}$ **3.** $\overline{CE}$ **4.** $\overline{DE}$

5. $\overline{AE}$ **6.** $\overline{FC}$ **7.** $\overline{GE}$ **8.** $\overline{BF}$

Refer to the coordinate plane at the right to find the coordinates of the midpoint of each segment.

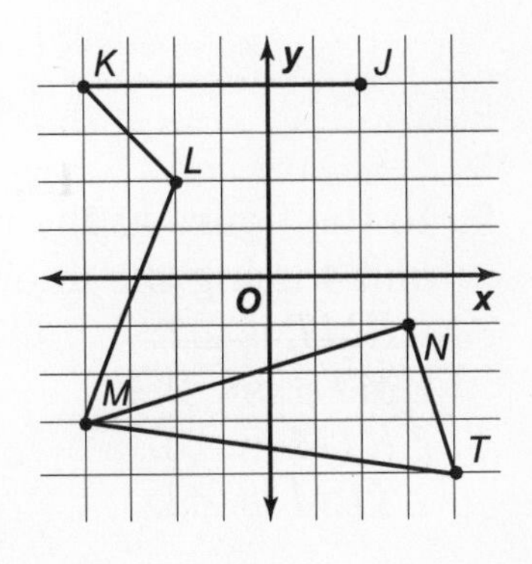

9. $\overline{JK}$ **10.** $\overline{KL}$

11. $\overline{LM}$ **12.** $\overline{MN}$

13. $\overline{NT}$ **14.** $\overline{MT}$

NAME ______________________ DATE ________ PERIOD ______

Study Guide

Angles

An **angle** is formed by two noncollinear rays with a common endpoint called a **vertex**. You could name the angle at the right as $\angle S$, $\angle RST$, $\angle TSR$, or $\angle 1$.

When two or more angles have a common vertex, you need to use either three letters or a number to name the angles. Make sure there is no doubt which angle your name describes.

Solve.

1. Label the three points K, N, and Q on the angle below so that the angle has sides $\overrightarrow{KQ}$ and $\overrightarrow{KN}$.

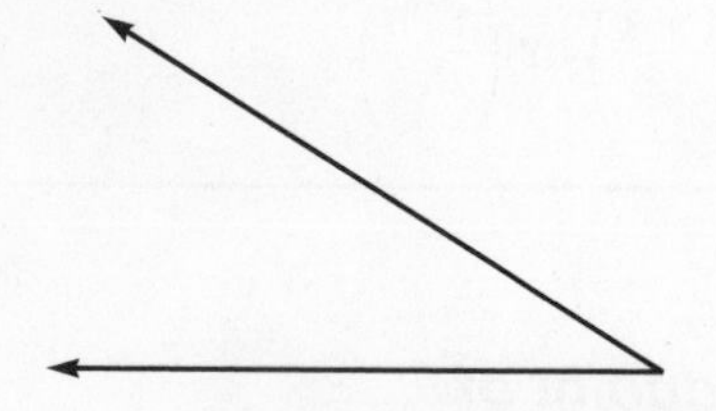

2. Name all angles having N as their vertex.

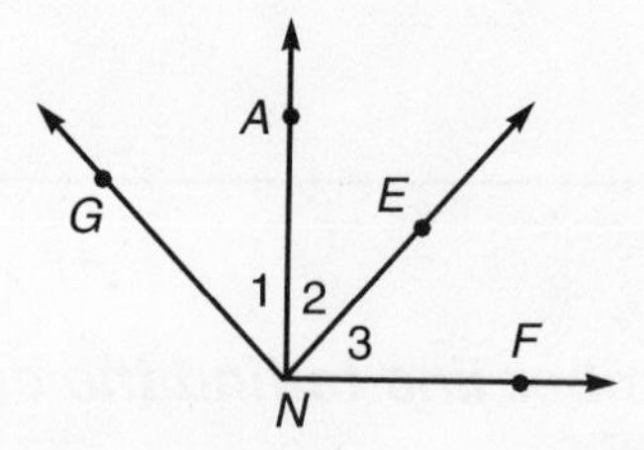

3. Draw $\angle BMA$ with sides $\overrightarrow{MB}$ and $\overrightarrow{MA}$ that has point P in its interior.

4. Using the figure from Exercise 3, draw $\overrightarrow{MP}$. Name the two new angles formed.

5. In the figure at the right, label angles 1, 2, 3, and 4 using the information below.
$\angle RMT$ is $\angle 1$.
$\angle MTS$ is $\angle 2$.
$\angle RTM$ is $\angle 3$.
$\angle TRM$ is $\angle 4$.

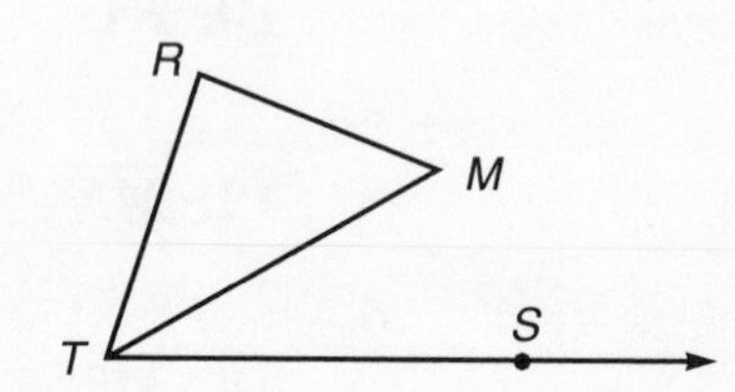

NAME ______________________ DATE ________ PERIOD ______

Study Guide

Student Edition
Pages 96–101

Angle Measure

Angles are measured in degrees.

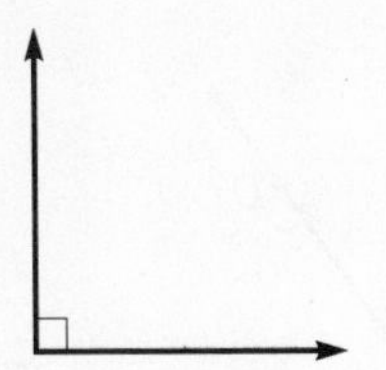

A **right angle** has a measure of 90.

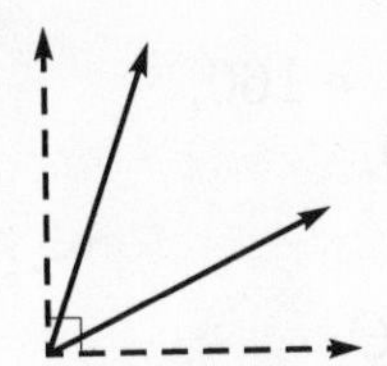

An **acute angle** has a measure between 0 and 90.

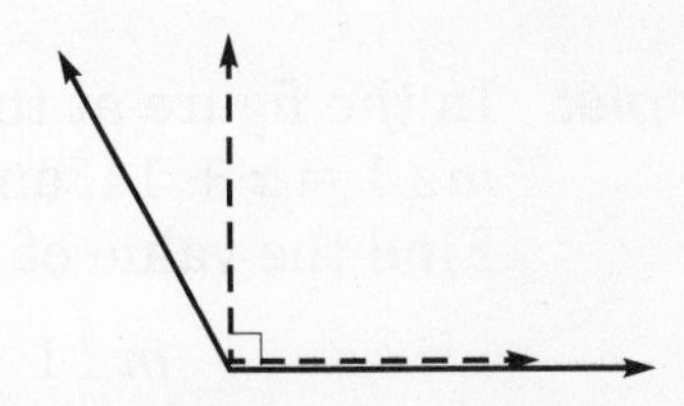

An **obtuse angle** has a measure between 90 and 180.

A **straight angle** has a measure of 180.

***Classify each angle as* acute, obtuse, *or* right.**

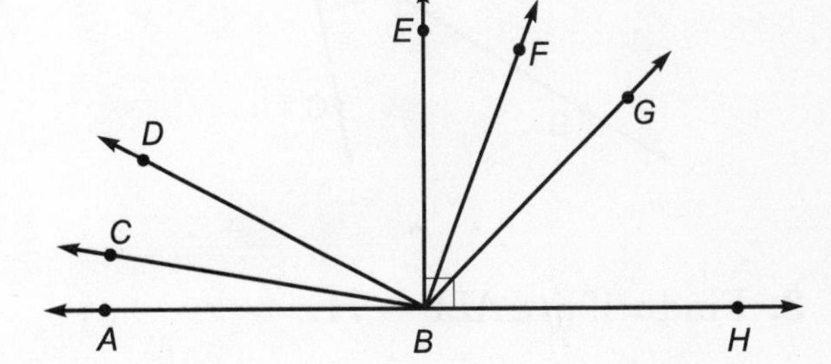

1. $\angle FBH$ **2.** $\angle CBD$

3. $\angle ABC$ **4.** $\angle ABG$

5. $\angle ABE$ **6.** $\angle EBH$

7. $\angle DBH$ **8.** $\angle FBG$

***Use a protractor to draw an angle having each measurement. Then classify each angle as* acute, obtuse, *or* right.**

9. 110° **10.** 28° **11.** 90°

NAME ______________________ DATE __________ PERIOD ______

Study Guide

Student Edition
Pages 104–109

The Angle Addition Postulate

According to the Angle Addition Postulate, if D is in the interior of $\angle ABC$, then $m\angle ABD + m\angle DBC = m\angle ABC$.

Example: In the figure at the right, $m\angle ABC = 160$, $m\angle 1 = x + 14$, and $m\angle 2 = 3x - 10$. Find the value of x.

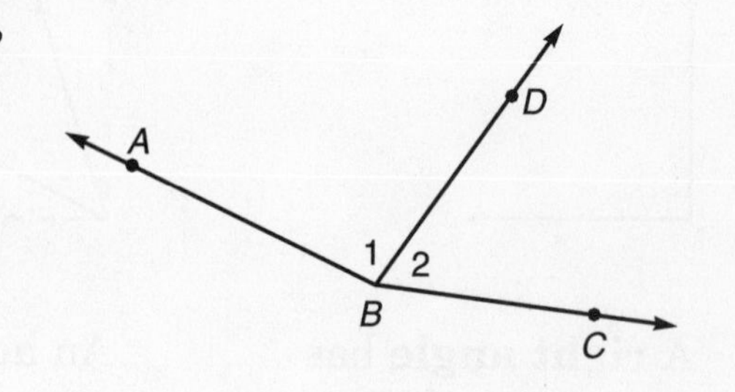

$$
\begin{aligned}
m\angle 1 + m\angle 2 &= m\angle ABC \\
(x + 14) + (3x - 10) &= 160 \\
4x + 4 &= 160 \\
4x &= 156 \\
x &= 39
\end{aligned}
$$

Solve.

1. Find $m\angle 1$ if $m\angle ARC = 78$.

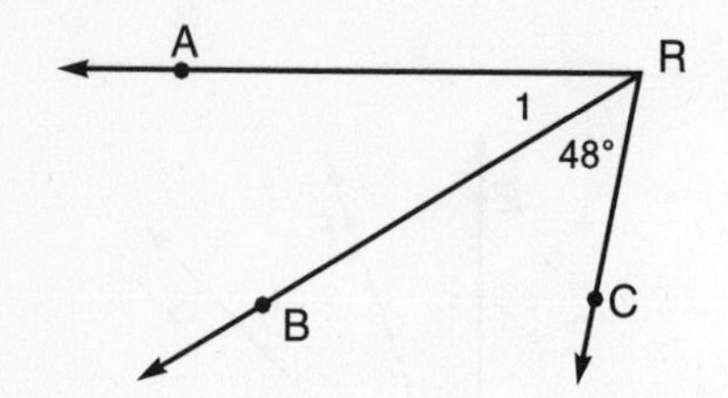

2. Find $m\angle 2$ if $m\angle YJK = 160$.

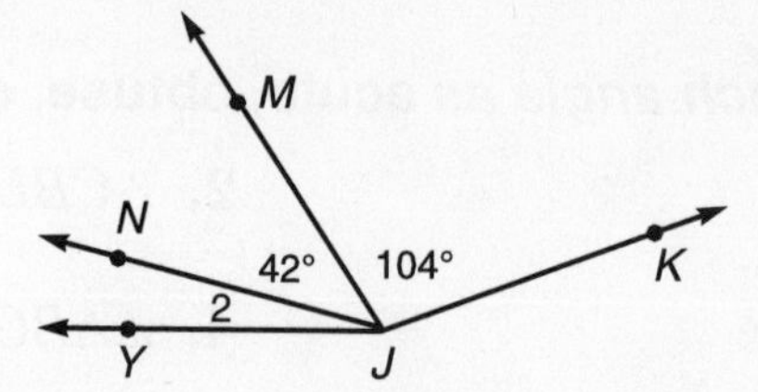

3. Find x if $m\angle ALY = 71$.

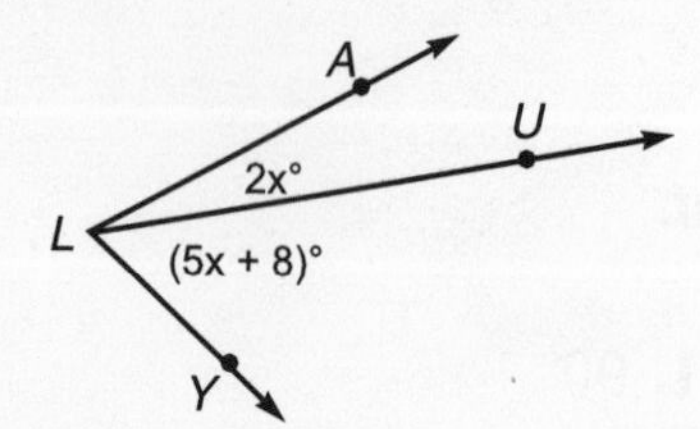

4. Find n if $m\angle QRS = 12n$.

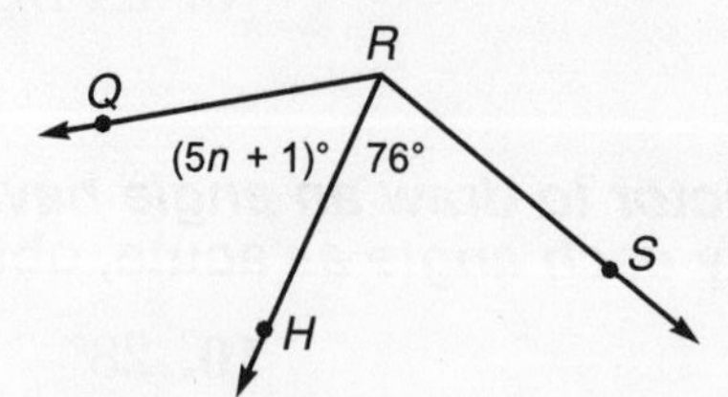

5. If $m\angle WYV = 4x - 2$, $m\angle VYZ = 2x - 5$, and $m\angle WYZ = 77$, find the measures of $\angle WYV$ and $\angle VYZ$.

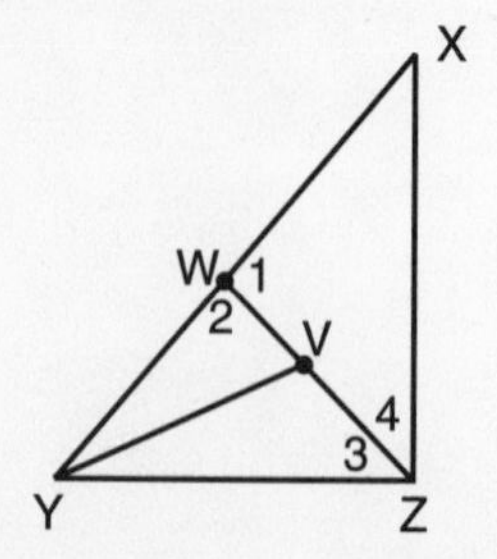

6. If $\overrightarrow{QS}$ bisects $\angle RQP$, $m\angle RQS = 2x + 10$, and $m\angle SQP = 3x - 18$, find $m\angle RQS$.

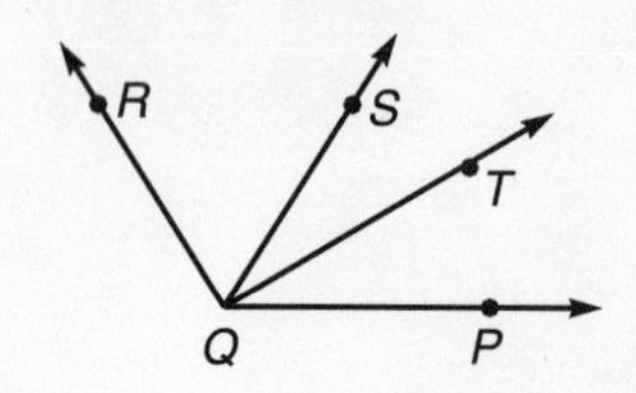

NAME ______________________ DATE __________ PERIOD ______

Study Guide

Student Edition
Pages 110–115

Adjacent Angles and Linear Pairs of Angles

Pairs of Angles		
Special Name	**Definition**	**Examples**
adjacent angles	angles in the same plane that have a common vertex and a common side, but no common interior points	3 4 ∠3 and ∠4 are adjacent angles.
linear pair	adjacent angles whose noncommon sides are opposite rays	5 6 ∠5 and ∠6 form a linear pair.

m∠1 = 45, m∠2 = 135, m∠3 = 125, m∠4 = 45, m∠5 = 135, m∠6 = 35, and ∠CAT is a right angle. Determine whether each statement is true or false.

1. ∠1 and ∠2 form a linear pair.
2. ∠4 and ∠5 form a linear pair.
3. ∠6 and ∠3 are adjacent angles.
4. ∠7 and ∠8 are adjacent angles.
5. ∠*CAT* and ∠7 are adjacent angles.

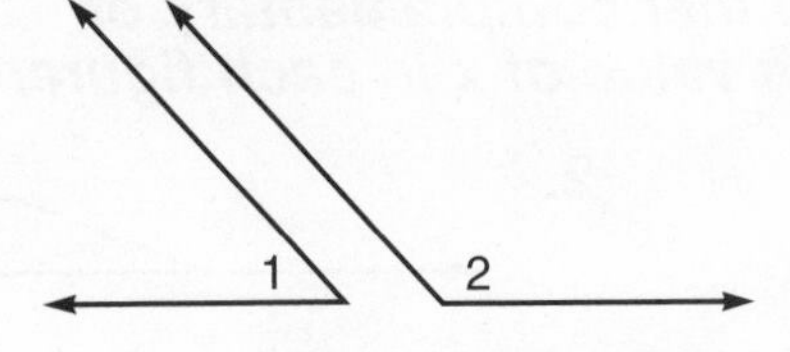

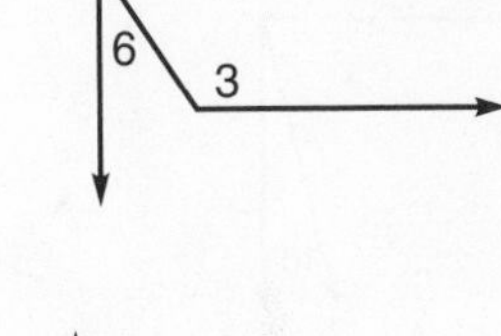

4 5

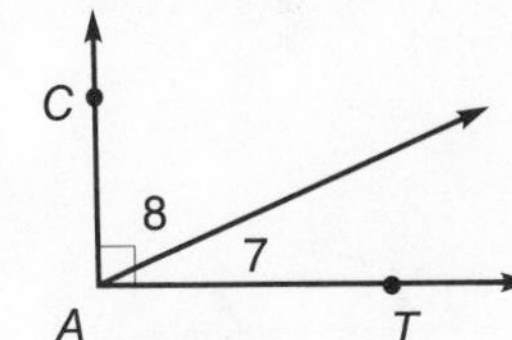

Use the terms* adjacent angles, linear pair, *or* neither *to describe angles 1 and 2 in as many ways as possible.

6.

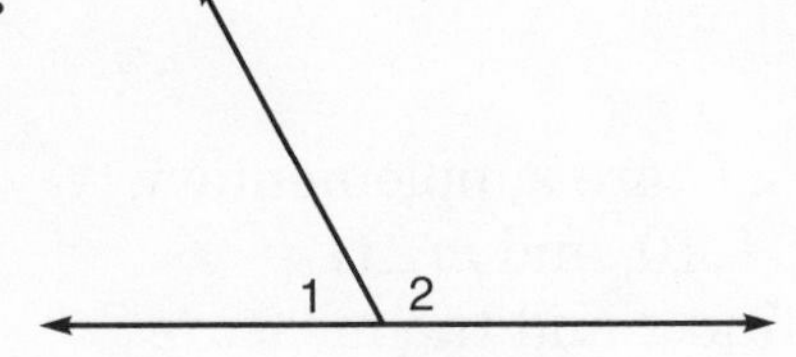

7.

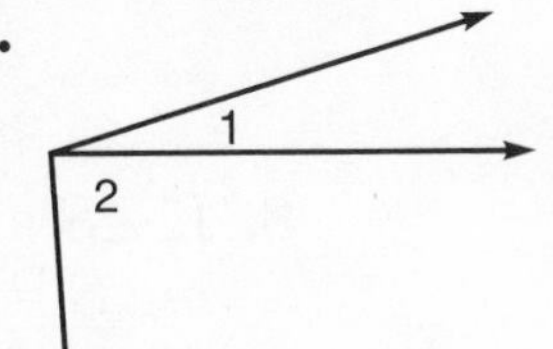

8.

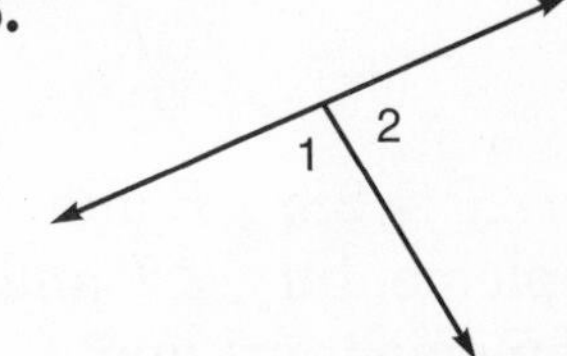

Study Guide

Complementary and Supplementary Angles

The table identifies several different types of angles that occur in pairs.

Pairs of Angles		
Special Name	**Definition**	**Examples**
complementary angles	two angles whose measures have a sum of 90	30° 60°
supplementary angles	two angles whose measures have a sum of 180	20° 160°

Each pair of angles is either complementary or supplementary. Find the value of x in each figure.

1. x° 15°

2.

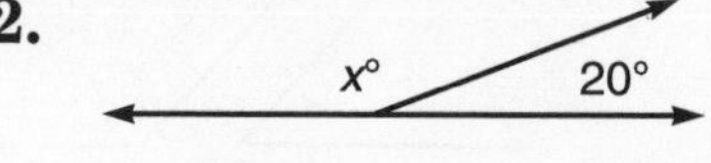

3.

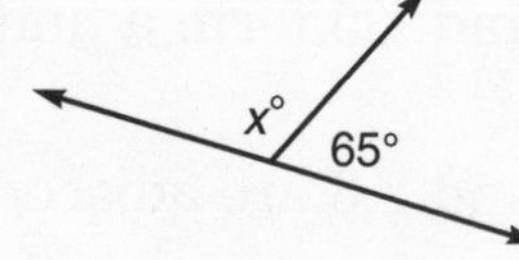

4.

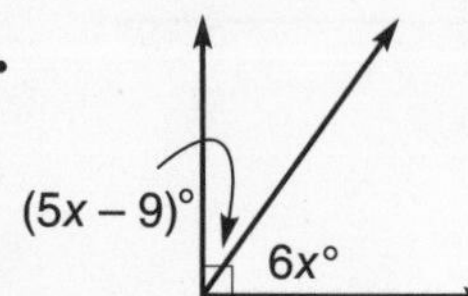

5.

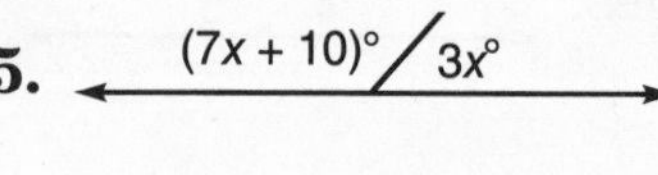

6.

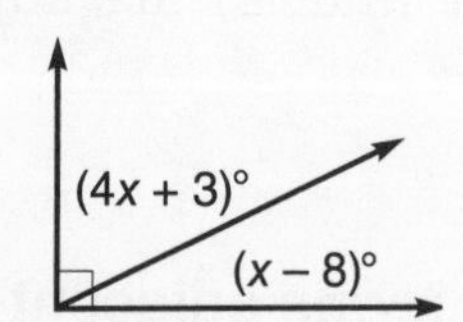

7. If $m\angle P = 28$, $\angle R$ and $\angle P$ are supplementary, $\angle T$ and $\angle P$ are complementary, and $\angle Z$ and $\angle T$ complementary, find $m\angle R$, $m\angle T$, and $m\angle Z$.

8. If $\angle S$ and $\angle G$ are supplementary, $m\angle S = 6x + 10$, and $m\angle G = 15x + 23$, find x and the measure of each angle.

NAME ______________________ DATE __________ PERIOD _____

Study Guide

Student Edition
Pages 122–127

Congruent Angles

Opposite angles formed by intersecting lines are called **vertical angles**. Vertical angles are always congruent. $\angle 1$ and $\angle 3$, and $\angle 2$ and $\angle 4$ are pairs of vertical angles.

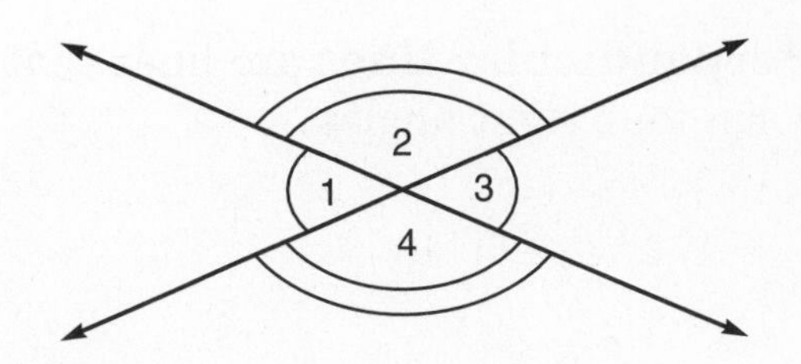

***Identify each pair of angles in Exercises 1–4 as* adjacent, vertical, complementary, supplementary, *and/or as a* linear pair.**

1. $\angle 1$ and $\angle 2$
2. $\angle 1$ and $\angle 4$

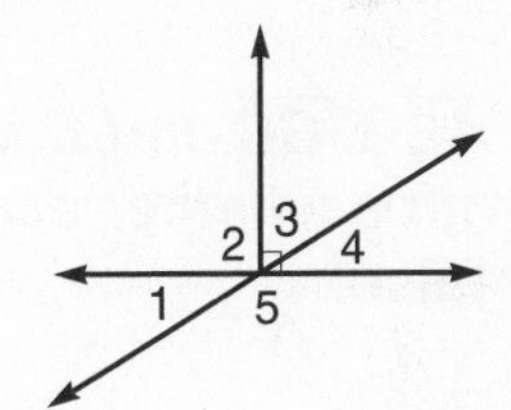

3. $\angle 3$ and $\angle 4$
4. $\angle 1$ and $\angle 5$

5. Find x, y, and z.

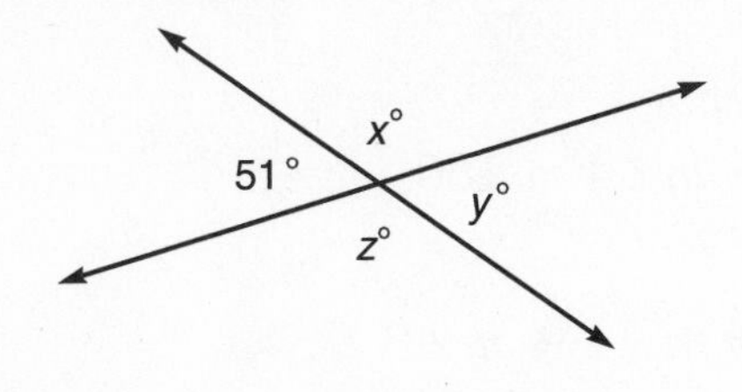

6. Find x and y if $\angle CBD \cong \angle FDG$.

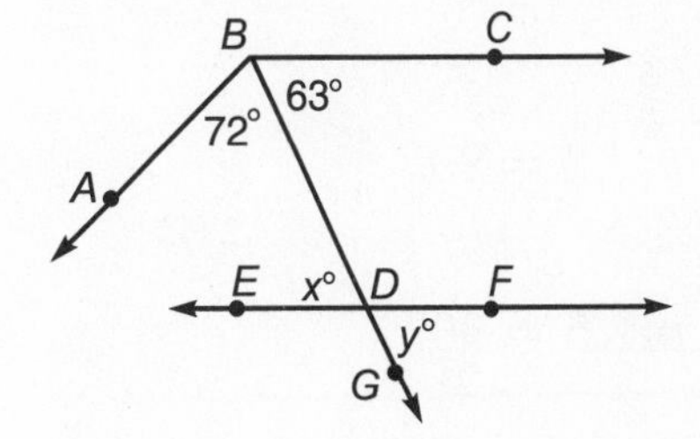

Use the figure shown to find each of the following.

7. x

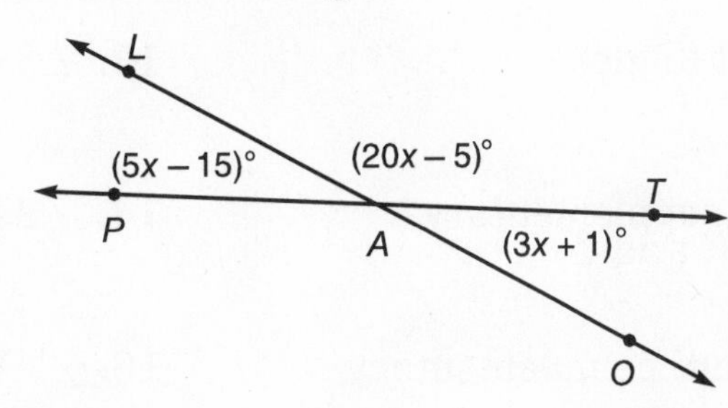

8. $m\angle LAT$

9. $m\angle TAO$

10. $m\angle PAO$

3-7

NAME ______ DATE ______ PERIOD ______

Study Guide

Student Edition
Pages 128–133

Perpendicular Lines

Perpendicular lines are lines that intersect to form four right angles.

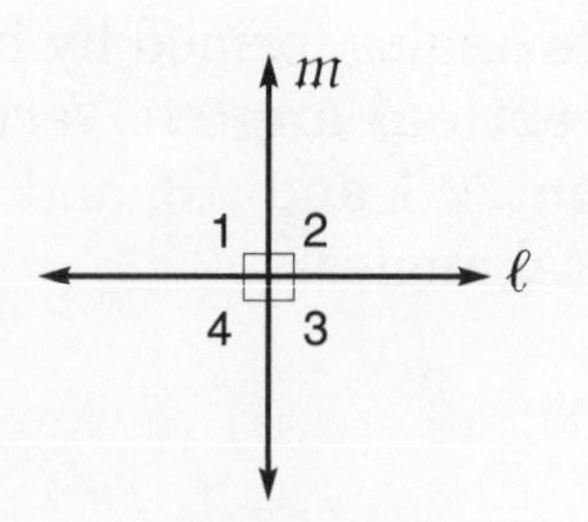

$\overleftrightarrow{AB} \perp \overleftrightarrow{FE}$, $\overleftrightarrow{AE} \perp \overleftrightarrow{GC}$ and C is the midpoint of $\overline{AE}$. Determine whether each of the following is true or false.

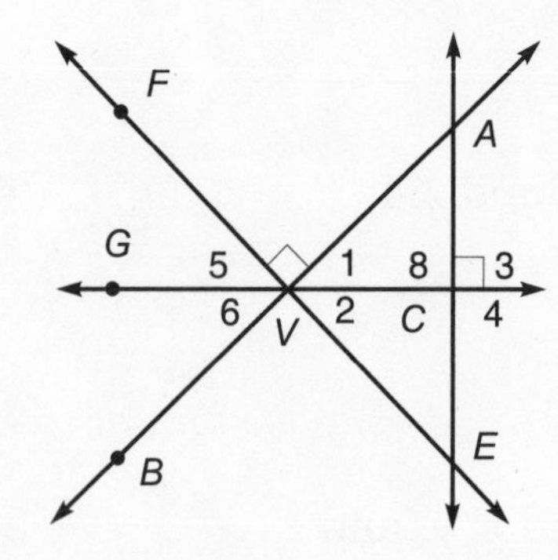

1. $\overleftrightarrow{GV} \perp \overleftrightarrow{AE}$

2. $\overleftrightarrow{AE} \perp \overleftrightarrow{FV}$

3. $\angle 4 \cong \angle 1$

4. $\angle 3 \cong \angle 4$

5. $m\angle 1 + m\angle 2 = 90$

6. $m\angle 3 + m\angle 4 = 180$

7. $m\angle 1 + m\angle 5 = 90$

8. $m\angle 4 = m\angle 1 + m\angle 2$

9. $m\angle AVF = 90$

10. $m\angle BVE = 90$

11. $\angle GVA$ is a right angle.

12. $\angle 3$, $\angle 4$, and $\angle 8$ are right angles.

13. $\angle 6$ and $\angle 3$ are supplementary.

14. $\angle 2$ and $\angle 6$ are complementary.

15. $\angle FVB$ and $\angle 4$ are complementary.

16. $\angle AVE$ and $\angle BVF$ are supplementary.

17. $\overleftrightarrow{AE}$ is the only line perpendicular to $\overleftrightarrow{GC}$ at C.

4-1

NAME ______________________ DATE ________ PERIOD ______

Study Guide

Student Edition
Pages 142–147

Parallel Lines and Planes

When planes do not intersect, they are said to be **parallel**. Also, when lines in the same plane do not intersect, they are parallel. But when lines are not in the same plane and do not intersect, they are **skew**.

Example: Name the parts of the triangular prism shown at the right. Sample answers are given.

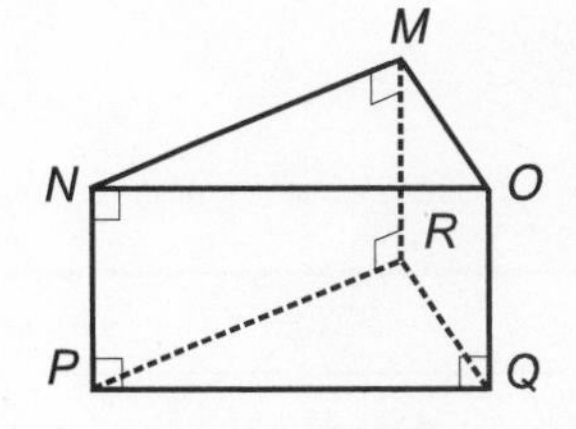

parallel planes: planes PQR and NOM
parallel segments: $\overline{MO}$ and $\overline{RQ}$
skew segments: $\overline{MN}$ and $\overline{RQ}$

Refer to the figure in the example.

1. Name two more pairs of parallel segments.

2. Name two more segments skew to $\overline{NM}$.

3. Name a segment that is parallel to plane MRQ.

Name the parts of the hexagonal prism shown at the right.

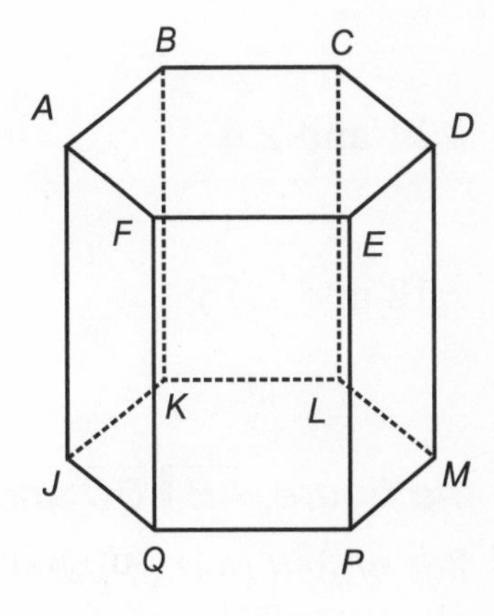

4. three segments that are parallel to $\overline{BC}$

5. three segments that are parallel to $\overline{JK}$

6. a segment that is skew to $\overline{QP}$

7. the plane that is parallel to plane AJQ

4-2

NAME ______________________ DATE ________ PERIOD _____

Study Guide

Student Edition
Pages 148–153

Parallel Lines and Transversals

A line that intersects two or more lines in a plane at different points is called a **transversal**. Eight angles are formed by a transversal and two lines.

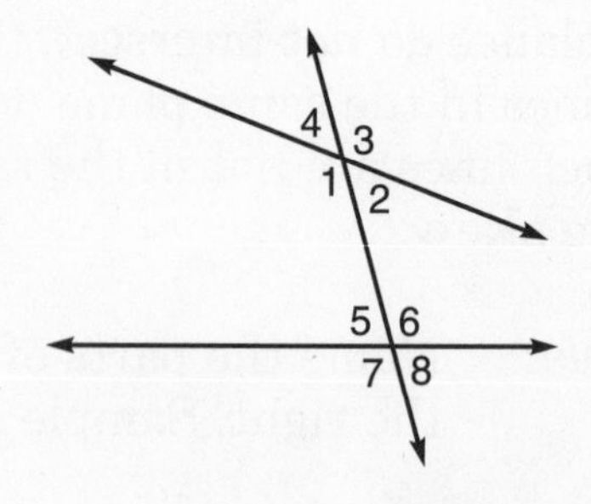

Types of Angles		
Angle	**Definition**	**Examples**
interior	lie between the two lines	$\angle 1$, $\angle 2$, $\angle 5$, $\angle 6$
alternate interior	on opposite sides of the transversal	$\angle 1$ and $\angle 6$, $\angle 2$ and $\angle 5$
consecutive interior	on the same side of the transversal	$\angle 1$ and $\angle 5$, $\angle 2$ and $\angle 6$
exterior	lie outside the two lines	$\angle 3$, $\angle 4$, $\angle 7$, $\angle 8$
alternate exterior	on opposite sides of the transversal	$\angle 3$ and $\angle 7$, $\angle 4$ and $\angle 8$

***Identify each pair of angles as* alternate interior, alternate exterior, consecutive interior, *or* vertical.**

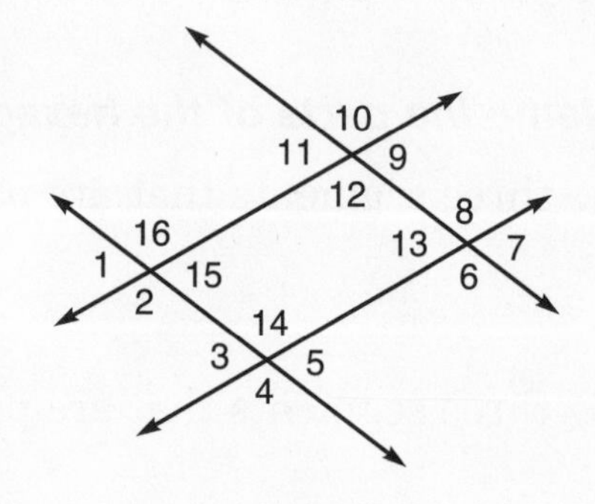

1. $\angle 6$ and $\angle 10$

2. $\angle 14$ and $\angle 13$

3. $\angle 14$ and $\angle 6$

4. $\angle 1$ and $\angle 5$

5. $\angle 12$ and $\angle 15$

6. $\angle 2$ and $\angle 16$

In the figure, $\overline{AB} \parallel \overline{DC}$ and $\overline{BC} \parallel \overline{AD}$.

7. For which pair of parallel lines are $\angle 1$ and $\angle 4$ alternate interior angles?

8. For which pair of parallel lines are $\angle 2$ and $\angle 3$ alternate interior angles?

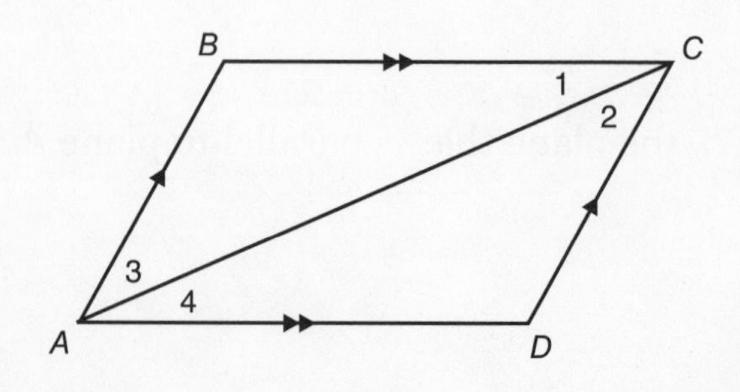

NAME ______________________ DATE ________ PERIOD _____

Study Guide

Student Edition
Pages 156–161

Transversals and Corresponding Angles

If two parallel lines are cut by a transversal, then the following pairs of angles are congruent.

corresponding angles
alternate interior angles
alternate exterior angles

If two parallel lines are cut by a transversal, then consecutive interior angles are supplementary.

Example: In the figure $m \parallel n$ and p is a transversal. If $m\angle 2 = 35$, find the measures of the remaining angles.

Since $m\angle 2 = 35$, $m\angle 8 = 35$ (corresponding angles).
Since $m\angle 2 = 35$, $m\angle 6 = 35$ (alternate interior angles).
Since $m\angle 8 = 35$, $m\angle 4 = 35$ (alternate exterior angles).

$m\angle 2 + m\angle 5 = 180$. Since consecutive interior angles are supplementary, $m\angle 5 = 145$, which implies that $m\angle 3$, $m\angle 7$, and $m\angle 1$ equal 145.

In the figure at the right $p \parallel q$, $m\angle 1 = 78$, and $m\angle 2 = 47$. Find the measure of each angle.

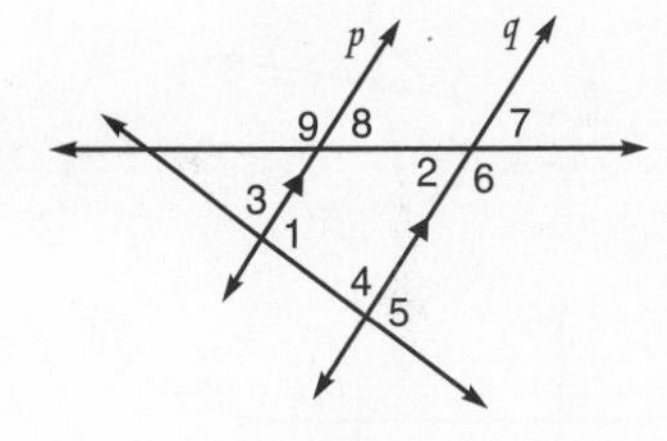

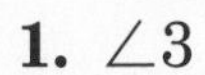

1. $\angle 3$ **2.** $\angle 4$ **3.** $\angle 5$

4. $\angle 6$ **5.** $\angle 7$ **6.** $\angle 8$ **7.** $\angle 9$

Find the values of x and y, in each figure.

8.

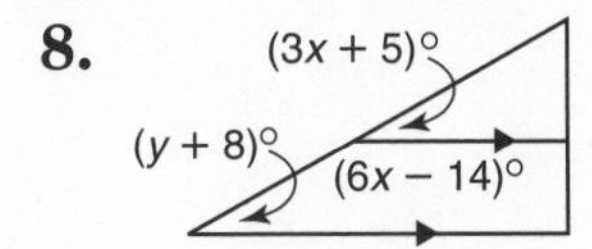

9.

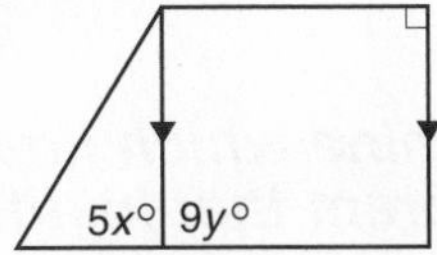

10.

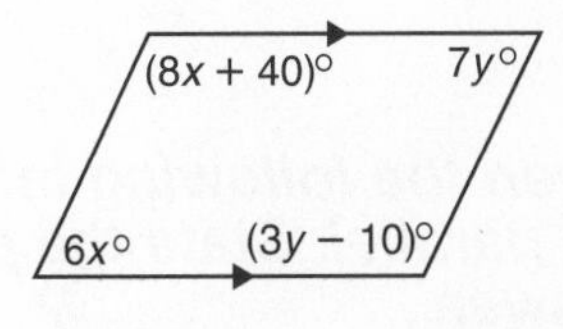

Find the values of x, y, and z in each figure.

11. (3z + 18)° x° 72° 3y°

12.

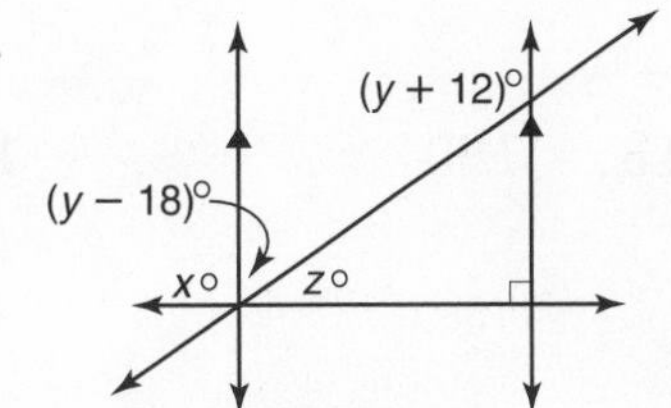

NAME ______________________ DATE ________ PERIOD ______

Study Guide

Student Edition
Pages 162–167

Proving Lines Parallel

Suppose two lines in a plane are cut by a transversal. With enough information about the angles that are formed, you can decide whether the two lines are parallel.

IF	THEN
corresponding angles are congruent, alternate interior angles are congruent, alternate exterior angles are congruent, consecutive interior angles are supplementary, the lines are perpendicular to the same line,	the lines are parallel.

Example: If $\angle 1 = \angle 2$, which lines must be parallel? Explain.

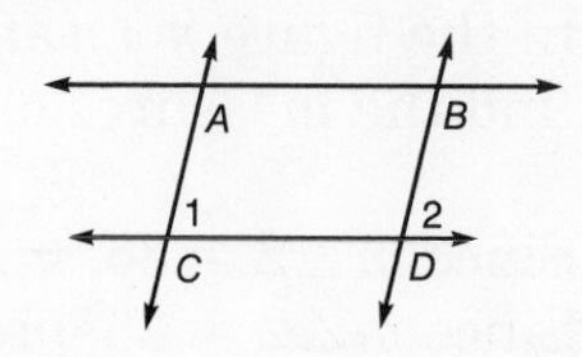

$\overleftrightarrow{AC} \parallel \overleftrightarrow{BD}$ because a pair of corresponding angles are congruent.

Find x so that $a \parallel b$.

1.

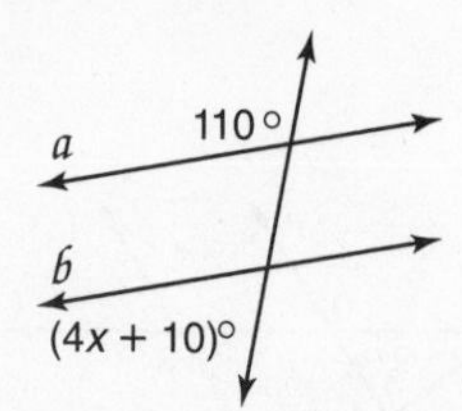

2.

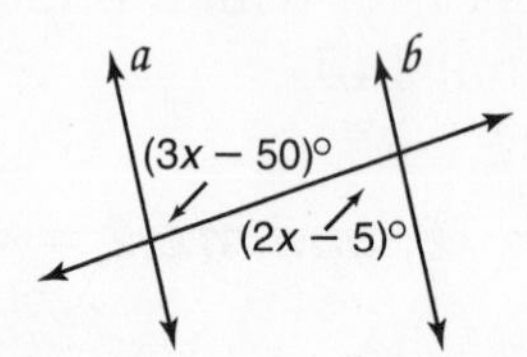

3.

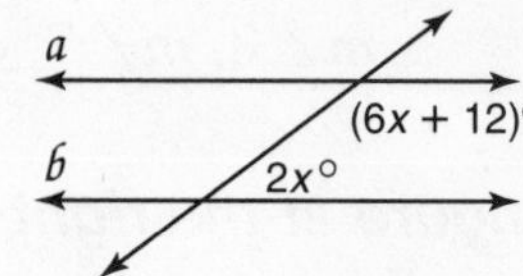

4.

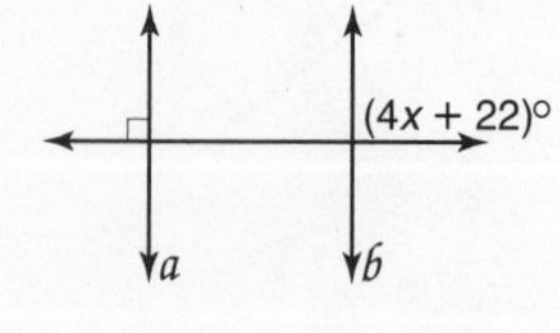

5.

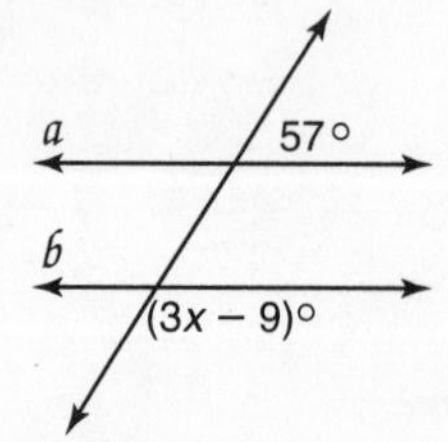

6.

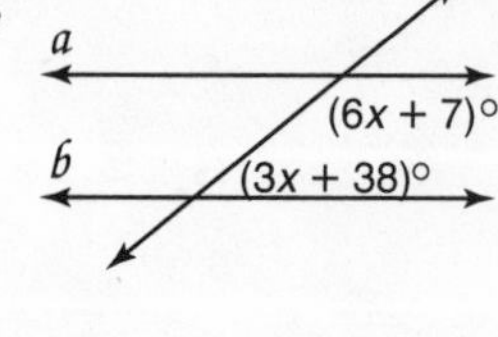

Given the following information, determine which lines, if any, are parallel. State the postulate or theorem that justifies your answer.

7. $\angle 1 \cong \angle 8$

8. $\angle 4 \cong \angle 9$

9. $m\angle 7 + m\angle 13 = 180$

10. $\angle 9 \cong \angle 13$

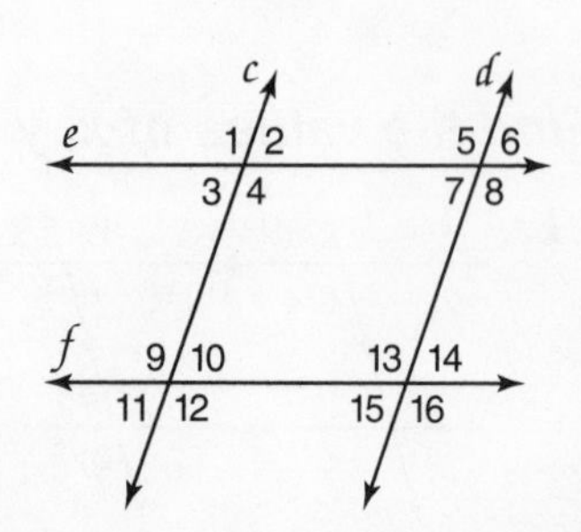

4-5 Study Guide

NAME ______________________ DATE __________ PERIOD ______

Student Edition
Pages 168–173

Slope

To find the slope of a line containing two points with coordinates (x_1, y_1) and (x_2, y_2), use the following formula.

$$m = \frac{y_2 - y_1}{x_2 - x_1}, \text{ where } x_1 \neq x_2$$

The slope of a vertical line, where $x_1 = x_2$, is undefined.

Two lines have the same slope if and only if they are parallel and nonvertical.

Two nonvertical lines are perpendicular if and only if the product of their slopes is −1.

Example: Find the slope of the line ℓ passing through $A(2, -5)$ and $B(-1, 3)$. State the slope of a line parallel to ℓ. Then state the slope of a line perpendicular to ℓ.

Let $(x_1, y_1) = (2, -5)$ and $(x_2, y_2) = (-1, 3)$.
Then $m = \frac{3 - (-5)}{-1 - 2} = -\frac{8}{3}$.

Any line in the coordinate plane parallel to ℓ has slope $-\frac{8}{3}$.

Since $-\frac{8}{3} \cdot \frac{3}{8} = -1$, the slope of a line perpendicular to the line ℓ is $\frac{3}{8}$.

Find the slope of the line passing through the given points.

1. $C(-2, -4)$, $D(8, 12)$

2. $J(-4, 6)$, $K(3, -10)$

3. $P(0, 12)$, $R(12, 0)$

4. $S(15, -15)$, $T(-15, 0)$

5. $F(21, 12)$, $G(-6, -4)$

6. $L(7, 0)$, $M(-17, 10)$

Find the slope of the line parallel to the line passing through each pair of points. Then state the slope of the line perpendicular to the line passing through each pair of points.

7. $I(9, -3)$, $J(6, -10)$

8. $G(-8, -12)$, $H(4, -1)$

9. $M(5, -2)$, $T(9, -6)$

4-6

NAME ______________________ DATE ________ PERIOD ______

Study Guide

Student Edition
Pages 174–179

Equations of Lines

You can write an equation of a line if you are given
- the slope and the coordinates of a point on the line, or
- the coordinates of two points on the line.

Example: Write the equation in slope-intercept form of the line that has slope 5 and an x-intercept of 3.

Since the slope is 5, you can substitute 5 for m in
$y = mx + b$.
$y = 5x + b$

Since the x-intercept is 3, the point (3, 0) is on the line.
$y = 5x + b$
$0 = 5(3) + b$ *$y = 0$ and $x = 3$*
$0 = 15 + b$
$b = -15$ *Solve for b.*
So the equation is $y = 5x - 15$.

If you know two points on a line, you will need to find the slope of the line passing through the points and then write the equation.

Write the equation in slope-intercept form of the line that satisfies the given conditions.

1. $m = 3$, y-intercept $= -4$

2. $m = -\frac{2}{5}$, x-intercept $= 6$

3. passes through (−5, 10) and (2, 4)

4. passes through (8, 6) and (−3, −3)

5. perpendicular to the y-axis, passes through (−6, 4)

6. parallel to the y-axis, passes through (−7, 3)

7. $m = 3$ and passes through (−4, 6)

8. perpendicular to the graph of $y = 4x - 1$ and passes through (6, −3)

5-1

NAME ______________________ DATE __________ PERIOD ______

Study Guide

Student Edition
Pages 188–192

Classifying Triangles

Triangles are classified in two different ways, either by their angles or by their sides.

Classification of Triangles			
Angles		**Sides**	
acute	all acute angles	scalene	no two sides congruent
obtuse	one obtuse angle	isosceles	at least two sides congrent
right	one right angle	equilateral	all sides congruent

Examples: Classify each triangle by its angles and by its sides.

1

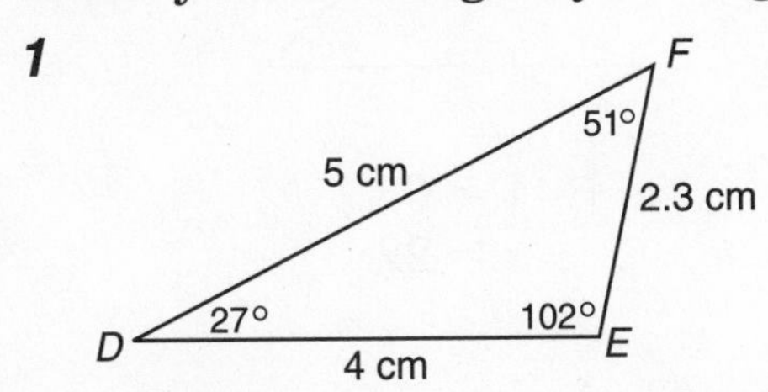

$\triangle DEF$ is obtuse and scalene.

2

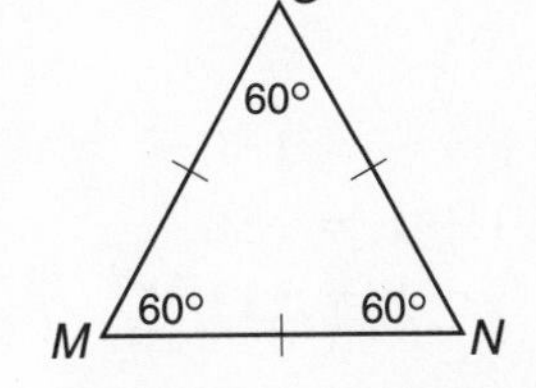

$\triangle MNO$ is acute and equilateral.

Use a protractor and ruler to draw triangles using the given conditions. If possible, classify each triangle by the measures of its angles and sides.

1. $\triangle KLM$, $m\angle K = 90$, $KL = 2.5$ cm, $KM = 3$ cm

2. $\triangle XYZ$, $m\angle X = 60$, $XY = YZ = ZX = 3$ cm

3. $\triangle DEF$, $m\angle D = 150$, $DE = DF = 1$ inch

4. $\triangle GHI$, $m\angle G = 30$, $m\angle H = 45$, $GH = 4$ cm

5. $\triangle NOP$, $m\angle N = 90$, $NO = NP = 2.5$ cm

6. $\triangle QRS$, $m\angle Q = 100$, $QS = 1$ inch, $QR = 1\frac{1}{2}$ inches

NAME ______________________________ DATE __________ PERIOD ______

5-2 Study Guide

Student Edition
Pages 193–197

Angles of a Triangle

On a separate sheet of paper, draw a triangle of any size. Label the three angles D, E, and F. Then tear off the three corners and rearrange them so that the three vertices meet at one point, with $\angle D$ and $\angle F$ each adjacent to $\angle E$. What do you notice?

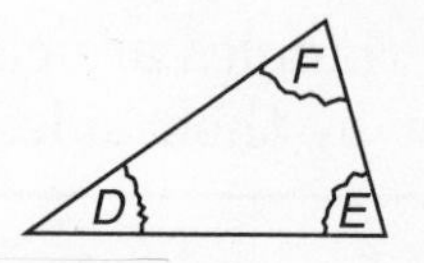

The sum of the measures of the angles of a triangle is 180.

Examples: Find the value of x in each figure.

1

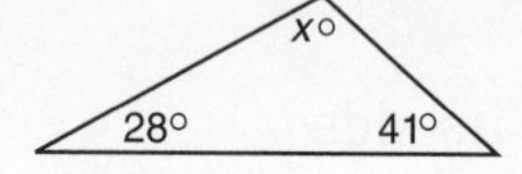

$$28 + 41 + x = 180$$
$$69 + x = 180$$
$$x = 111$$

2

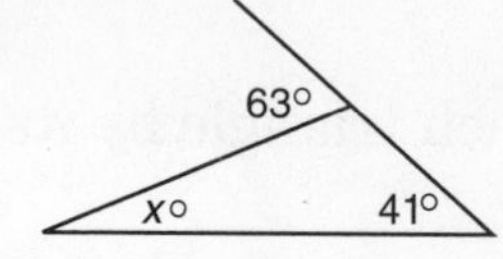

$$x + 41 = 63$$
$$x = 22$$

Find the value of x.

1.

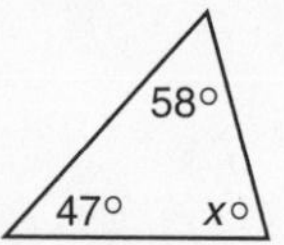

2.

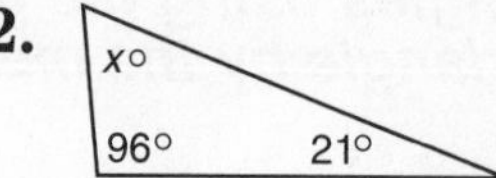

3.

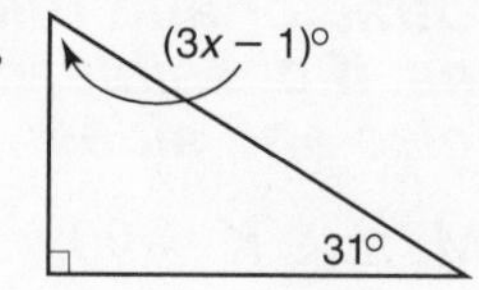

4.

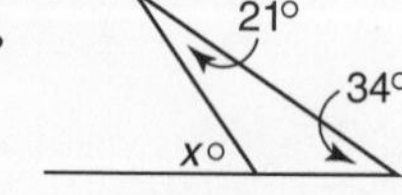

5.

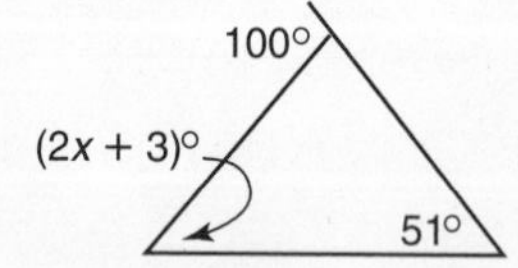

6.

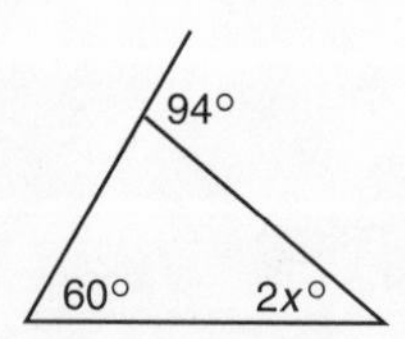

Find the measure of each angle.

7. $\angle 1$ **8.** $\angle 2$ **9.** $\angle 3$

10. $\angle 4$ **11.** $\angle 5$ **12.** $\angle 6$

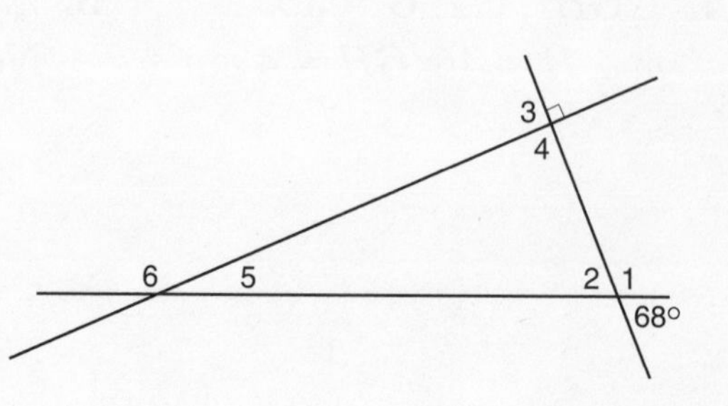

NAME ______________________ DATE __________ PERIOD ______

Study Guide

Student Edition
Pages 198–202

Geometry in Motion

Transformations		
Term	**Definition**	**Examples**
translation	slide a figure from one position to another without turning it	
reflection	flip a figure over a line	
rotation	turn a figure around a fixed point	

Determine whether each figure is a* translation, reflection, *or* rotation *of the given figure.

1. 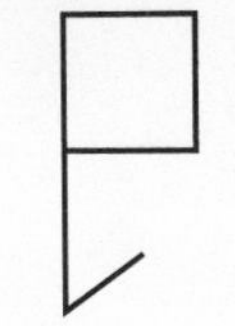

a. 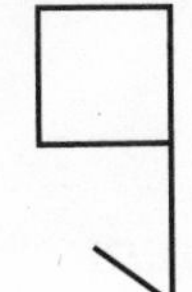

b. 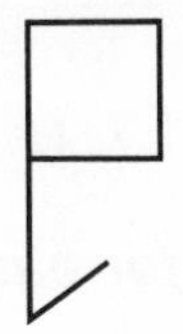

c. 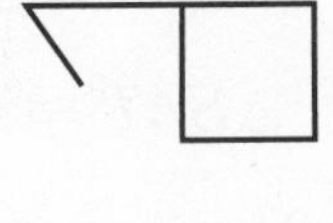

2. 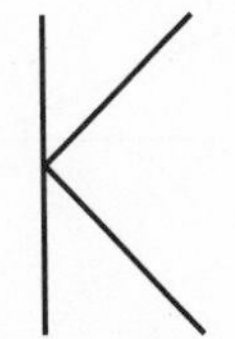

a. 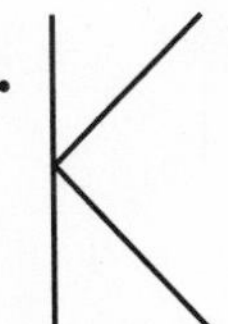

b. 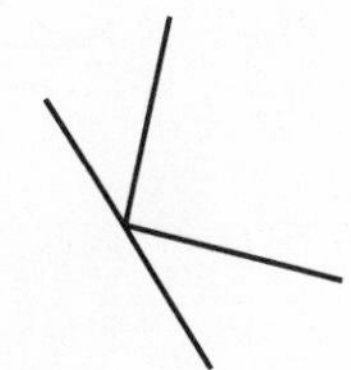

c. 

3. 

a. 

b. 

c. 

5-4

NAME ________________________ DATE __________ PERIOD ______

Study Guide

Student Edition
Pages 203–207

Congruent Triangles

When two figures have exactly the same shape and size, they are said to be congruent. For two congruent triangles there are three pairs of corresponding (matching) sides and three pairs of corresponding angles. To write a correspondence statement about congruent triangles, you should name corresponding angles in the same order. Remember that congruent parts are marked by identical markings.

Example: Write a correspondence statement for the triangles in the diagram.

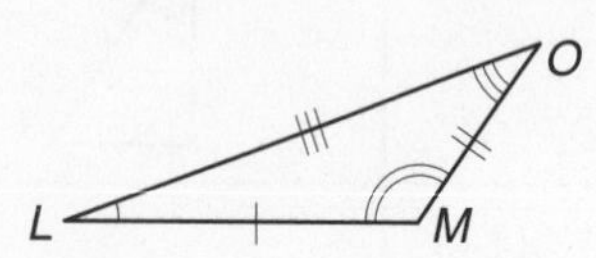

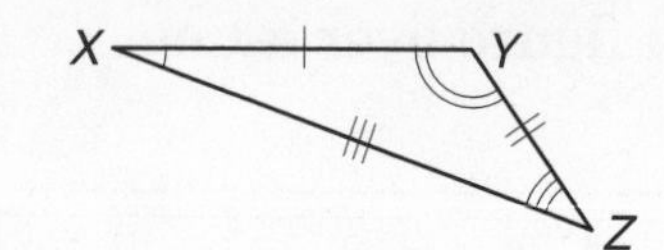

$\triangle LMO \cong \triangle XYZ$

Complete each correspondence statement.

1.

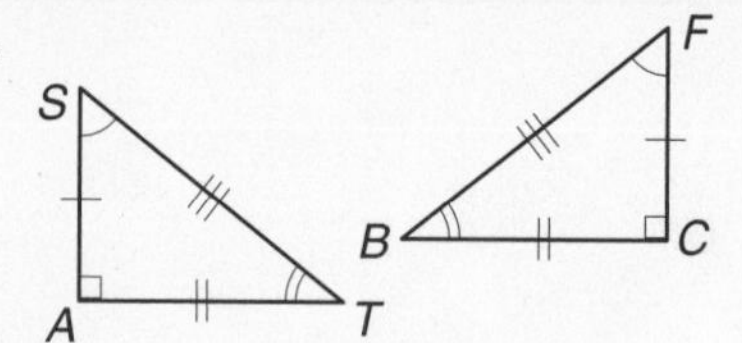

$\triangle SAT \cong \triangle$?

2.

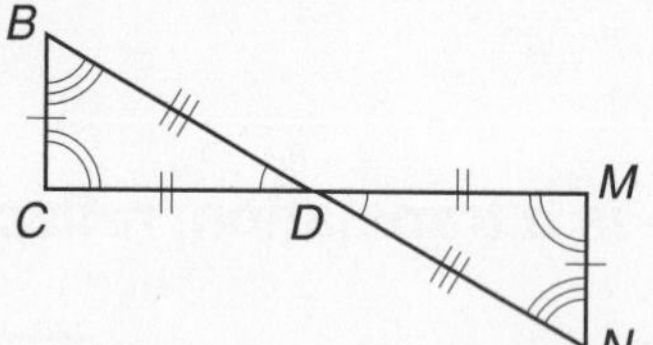

$\triangle BCD \cong \triangle$?

3.

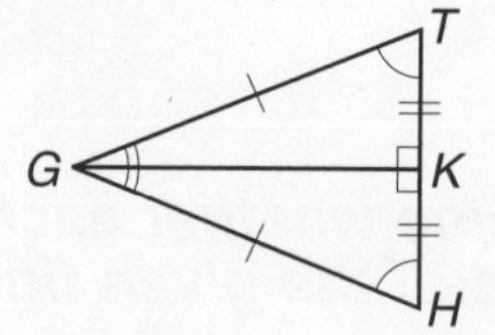

$\triangle GHK \cong \triangle$?

Write a congruence statement for each pair of congruent triangles.

4.

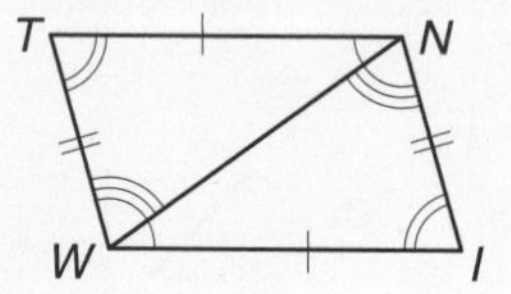

5.

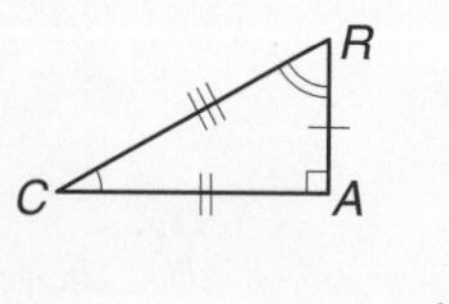

6.

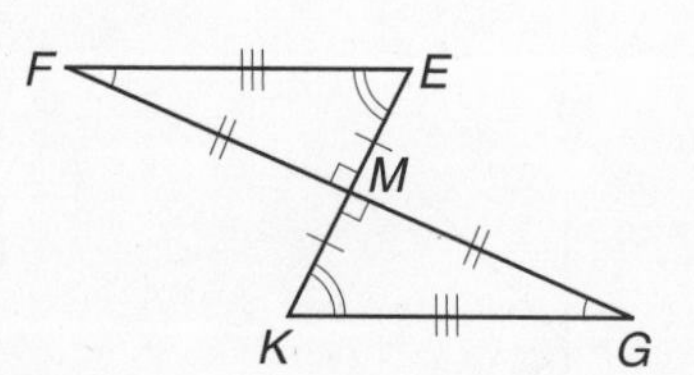

Draw triangles △EDG and △QRS. Label the corresponding parts if △EDG ≅ △QRS. Then complete each statement.

7. $\angle E \cong$?

8. $\overline{DG} \cong$?

9. $\angle EDG \cong$?

10. $\overline{GE} \cong$?

11. $\overline{ED} \cong$?

12. $\angle EGD \cong$?

NAME ______________________ DATE __________ PERIOD ______

Study Guide

Student Edition
Pages 210–214

SSS and SAS

You can show that two triangles are congruent with the following.

SSS Postulate (side-side-side)	If three sides of one triangle are congruent to three corresponding sides of another triangle, the triangles are congruent.
SAS Postulate (side-angle-side)	If two sides and the included angle of one triangle are congruent to the corresponding sides and included angle of another triangle, the triangles are congruent.

Determine whether each pair of triangles is congruent. If so, write a congruence statement and explain why the triangles are congruent.

1.

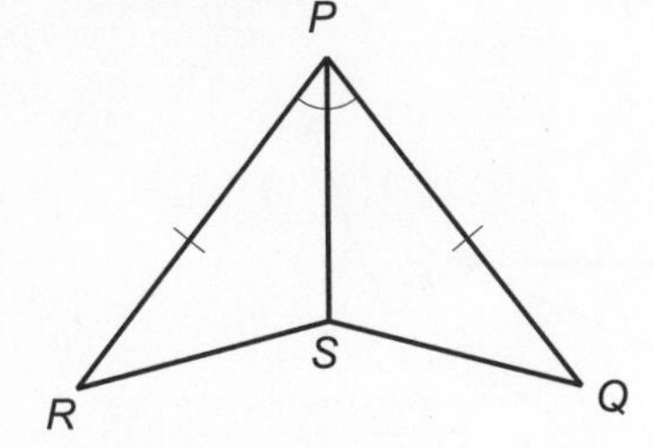

2.

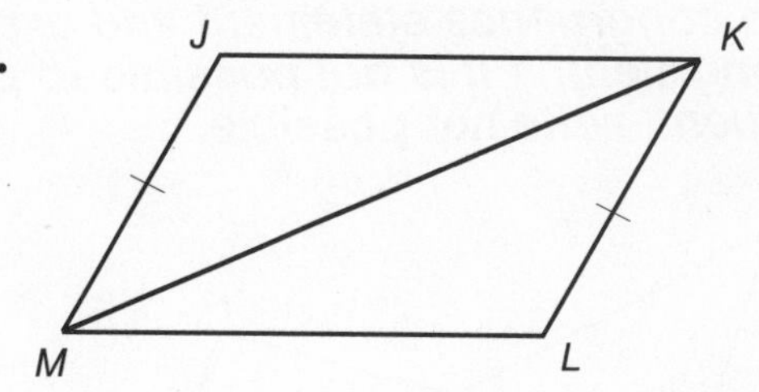

3.

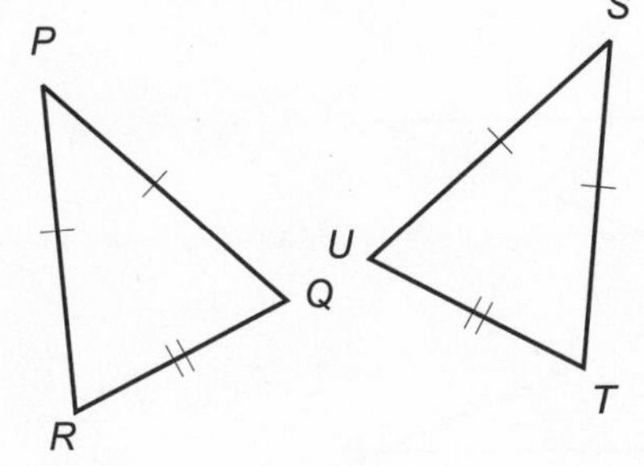

4.

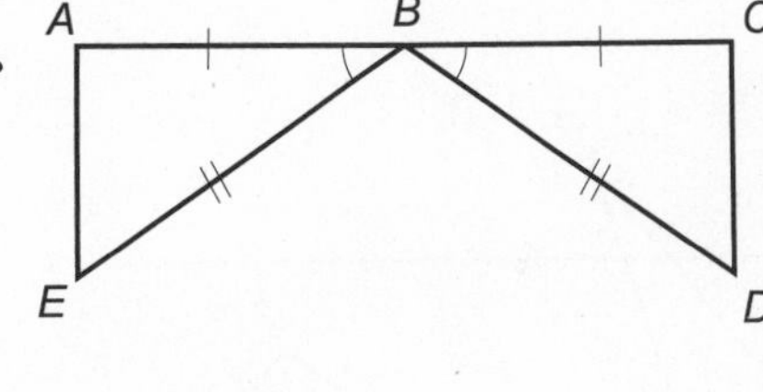

5.

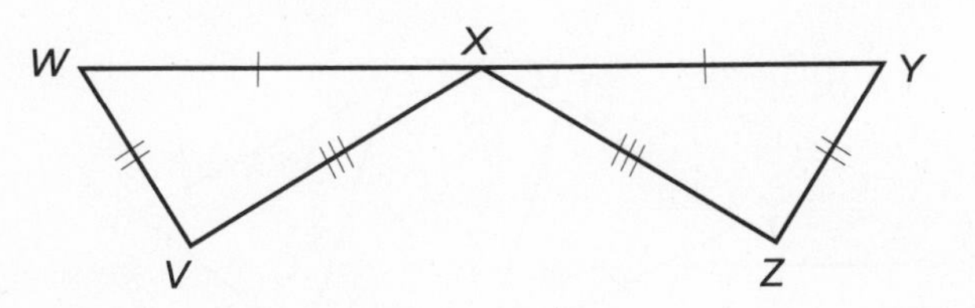

6.

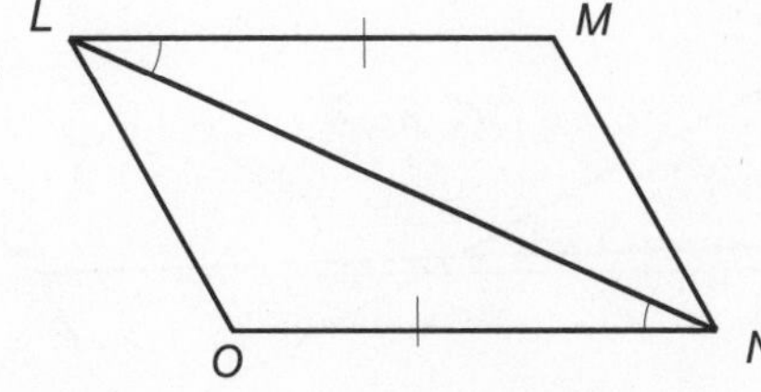

NAME ______________________________ DATE __________ PERIOD ______

Study Guide

Student Edition
Pages 215–219

ASA and AAS

You can show that two triangles are congruent with the following.

ASA Postulate (angle-side-angle)	If two angles and the included side of one triangle are congruent to the corresponding angles and included side of another triangle, the triangles are congruent.
AAS Theorem (angle-angle-side)	If two angles and a nonincluded side of one triangle are congruent to the corresponding two angles and nonincluded side of another triangle, the triangles are congruent.

***Determine whether each pair of triangles is congruent. If so, write a congruence statement and explain why the triangles are congruent. If it is not possible to prove that they are congruent, write* not possible.**

1.

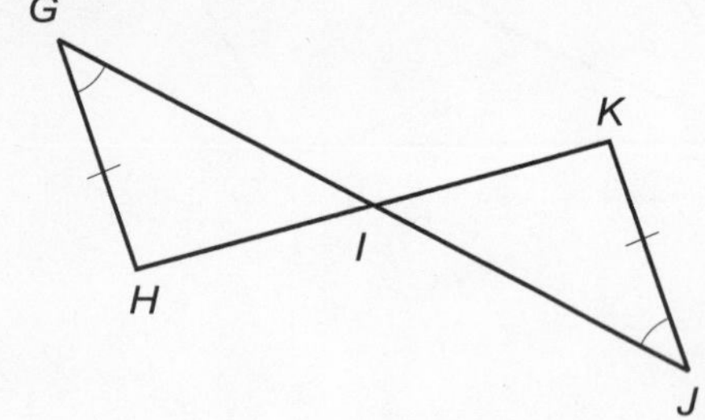

2.

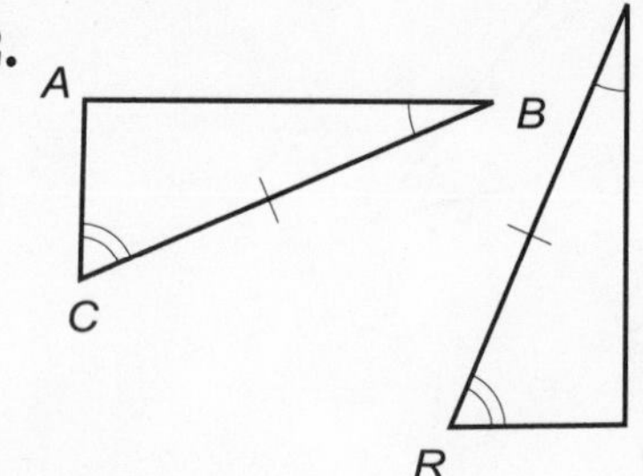

3.

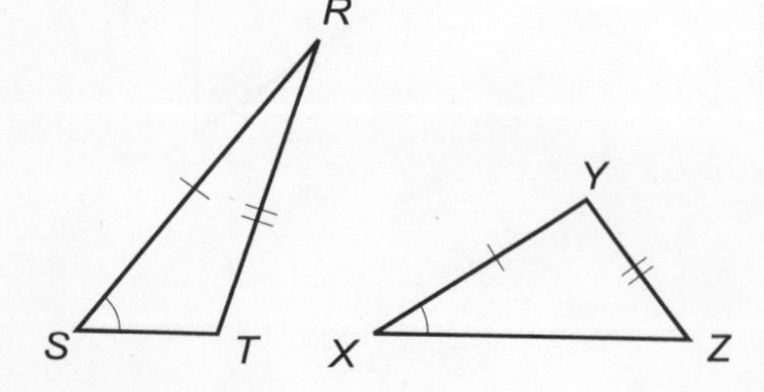

4.

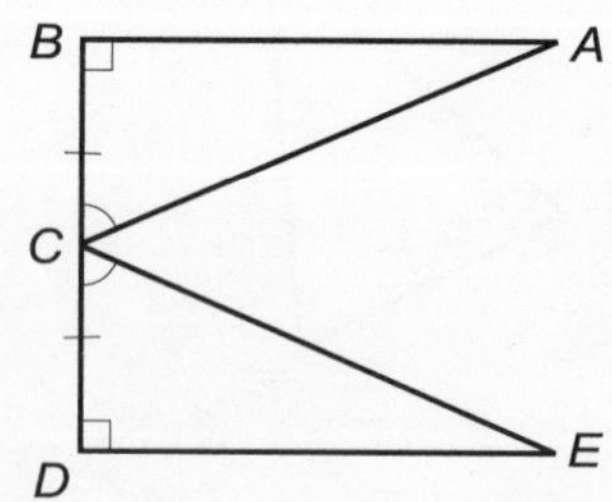

5.

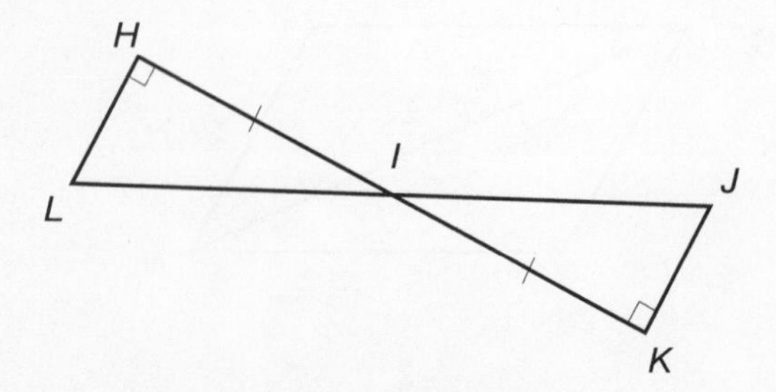

6.

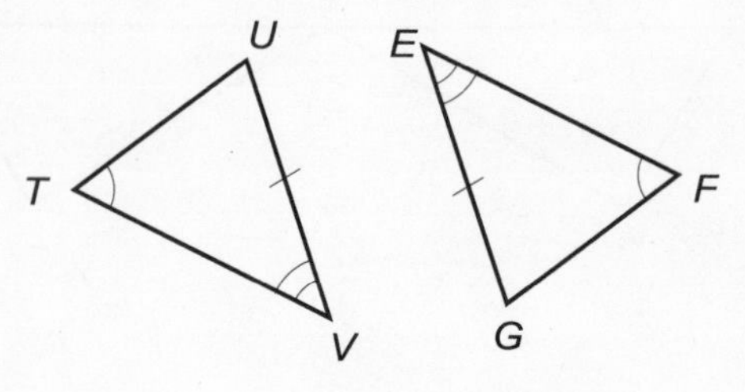

NAME ______________________ DATE ________ PERIOD ______

Study Guide

Student Edition
Pages 228–233

Medians

A **median** is a segment that joins a vertex of a triangle and the midpoint of the side opposite that vertex. The medians of a triangle intersect at a common point called the **centroid**. An important theorem about medians and centroids is as follows.

The length of the segment from the vertex to the centroid is twice the length of the segment from the centroid to the midpoint.

Example:

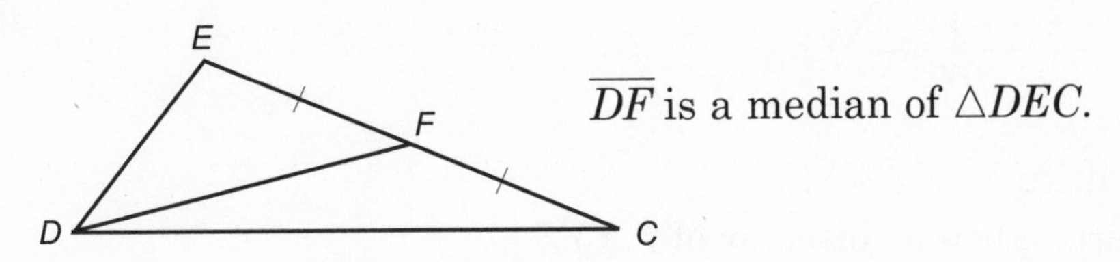

$\overline{DF}$ is a median of $\triangle DEC$.

In $\triangle ABC$, $\overline{AN}$, $\overline{BP}$, and $\overline{CM}$ are medians.

1. If $BP = 10$, find BE.

2. If $EM = 3$, find EC.

3. If $EN = 12$, find AN.

4. If $CM = 3x + 6$ and $CE = x + 12$, what is x?

5. If $EN = x - 5$ and $AE = x + 17$, find AN.

Draw and label a figure to illustrate each situation.

6. $\triangle NRW$ is a right triangle with right angle at N. $\overline{NX}$ is a median of $\triangle NRW$.

7. $\overline{OQ}$ is a median of $\triangle POM$.

NAME ______________________________ DATE __________ PERIOD ______

Study Guide

Student Edition
Pages 234–239

Altitudes and Perpendicular Bisectors

An **altitude** of a triangle is a perpendicular segment that has one endpoint at a vertex of the triangle and the other endpoint on the side opposite that vertex. A **perpendicular bisector** of a side of a triangle is a segment or line that contains the midpoint of that side and is perpendicular to that side.

Example:

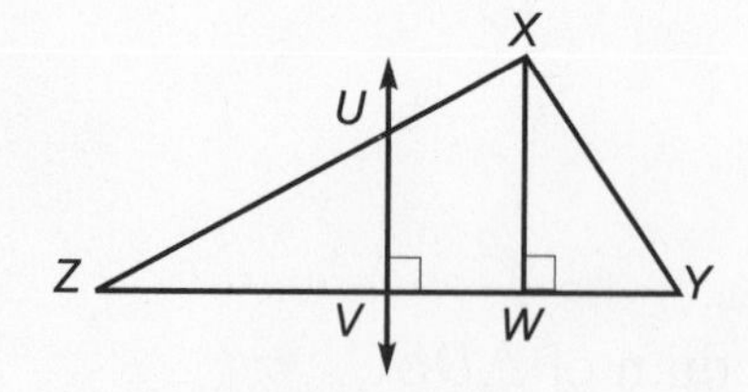

$\overline{XW}$ is an altitude of $\triangle XYZ$.
$\overline{UV}$ is a perpendicular bisector of $\triangle XYZ$.

In $\triangle FGH$, $\overline{GC} \cong \overline{CH}$, $\overline{FE} \cong \overline{EH}$, and $\angle FGD \cong \angle HGD$.

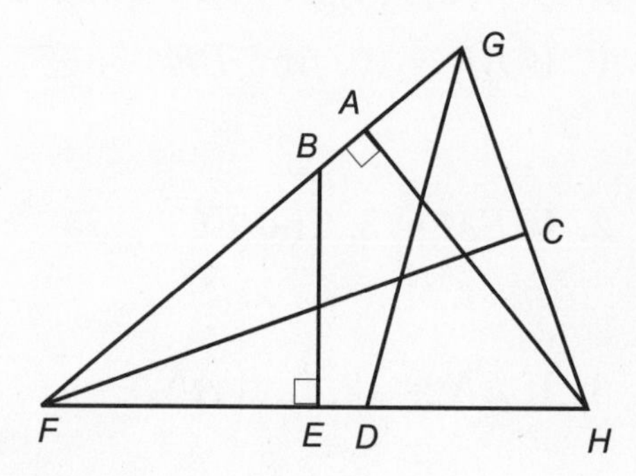

1. Which segment is a median of $\triangle FGH$?

2. Which segment is an altitude of $\triangle FGH$?

3. Which segment is a perpendicular bisector of $\triangle FGH$?

Draw and label a figure to illustrate each situation.

4. $\overline{BV}$ is a median and a perpendicular bisector of $\triangle BST$.

5. $\overline{KT}$ is an altitude of $\triangle KLM$, and L is between T and M.

NAME ______________________ DATE __________ PERIOD _____

Study Guide

Angle Bisectors of Triangles

An **angle bisector** of a triangle is a segment that bisects an angle of the triangle and has one endpoint at the vertex of that angle and the other endpoint on the side opposite that vertex.

Example:

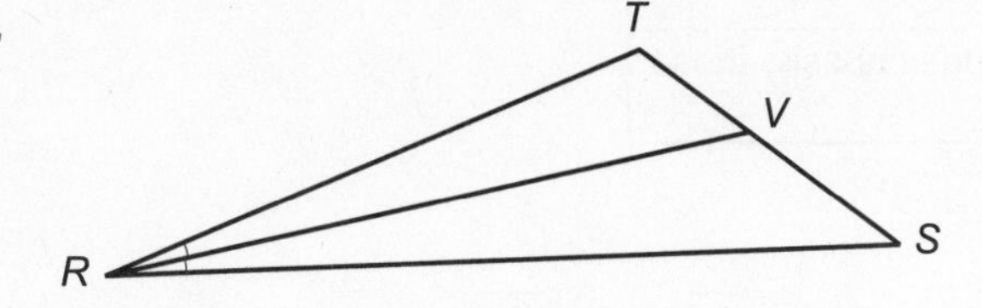

$\overline{RV}$ is an angle bisector of $\triangle RST$.

In $\triangle ACE$, $\overline{CF}$, $\overline{EB}$, and $\overline{AD}$ are angle bisectors.

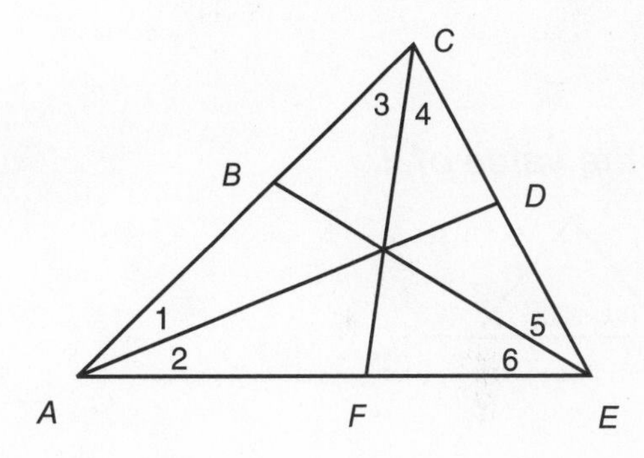

1. $m\angle 3 =$ ________

2. $m\angle 2 =$ ________

3. ________ bisects $\angle CAE$.

4. ________ bisects $\angle ACE$.

5. $m\angle 6 =$ ________$(m\angle CEA)$

6. $m\angle ACE =$ ________$(m\angle 3)$

7. Draw and label a figure to illustrate this situation. $\overline{HS}$ is an angle bisector of $\triangle GHI$, and S is between G and I.

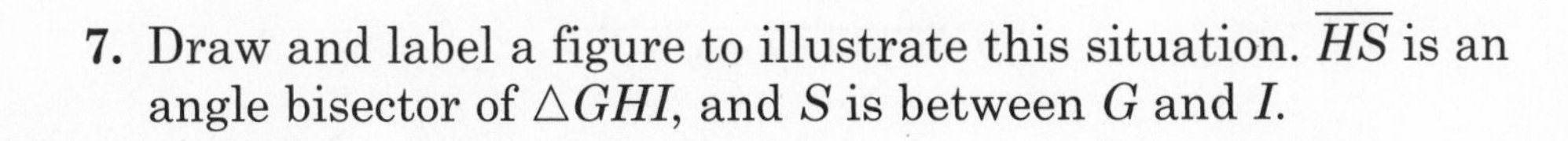

Study Guide

Student Edition
Pages 246–250

Isosceles Triangles

Remember that two sides of an isosceles triangle are congruent. Two important theorems about isosceles triangles are as follows.

If two sides of a triangle are congruent, then the angles opposite those sides are congruent.
If two angles of a triangle are congruent, then the sides opposite those angles are congruent.

Example: Find the value of x.

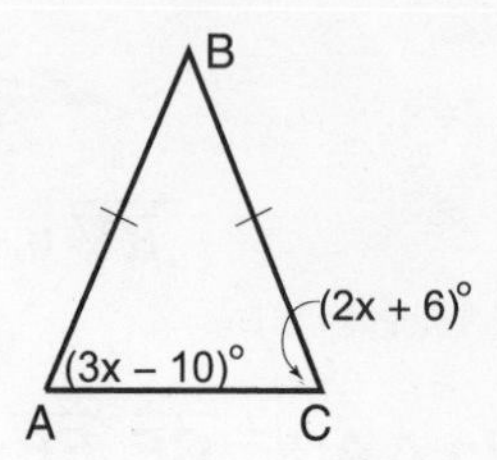

Since $\overline{AB} \cong \overline{BC}$, the angles opposite $\overline{AB}$ and $\overline{BC}$ are congruent. So $m\angle A = m\angle C$.
Therefore, $3x - 10 = 2x + 6$.

$$3x - 10 = 2x + 6$$
$$3x - 10 + 10 = 2x + 6 + 10 \quad \textit{Add 10 to each side.}$$
$$3x = 2x + 16$$
$$3x - 2x = 2x + 16 - 2x \quad \textit{Subtract 2x from each side.}$$
$$x = 16$$

Find the value of x.

1.

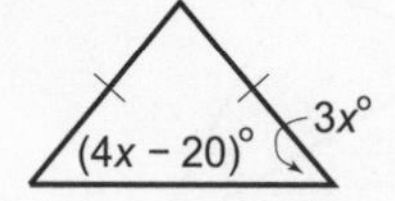

2.

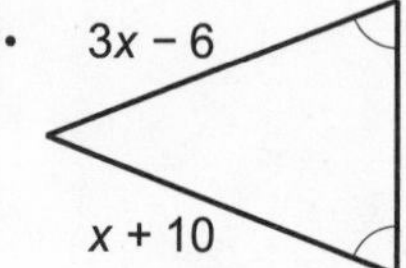

3.

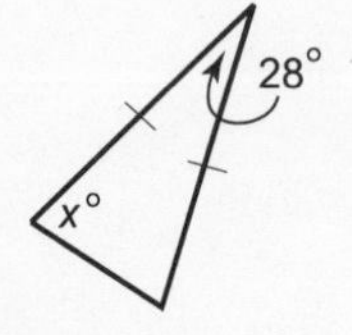

4.

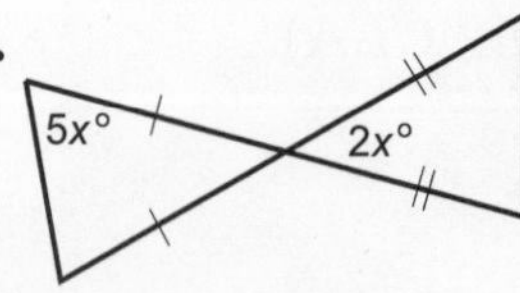

5.

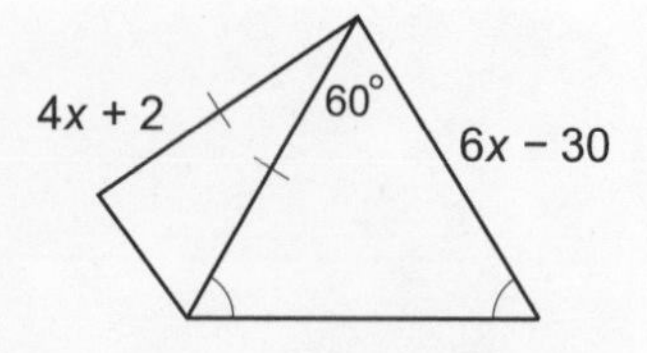

6.

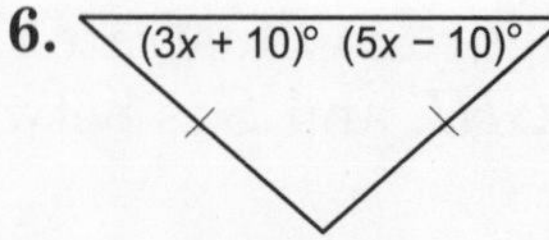

NAME ______________________ DATE __________ PERIOD ______

Study Guide

Student Edition
Pages 251–255

Right Triangles

Two right triangles are congruent if one of the following conditions exist.

Theorem 6-6 LL	If two legs of one right triangle are congruent to the corresponding legs of another right triangle, then the triangles are congruent.
Theorem 6-7 HA	If the hypotenuse and an acute angle of one right triangle are congruent to the hypotenuse and corresponding angle of another right triangle, then the two triangles are congruent.
Theorem 6-8 LA	If one leg and an acute angle of a right triangle are congruent to the corresponding leg and angle of another right triangle, then the triangles are congruent.
Postulate 6-1 HL	If the hypotenuse and a leg of one right triangle are congruent to the hypotenuse and corresponding leg of another right triangle, then the triangles are congruent.

State the additional information needed to prove each pair of triangles congruent by the given theorem or postulate.

1. HL

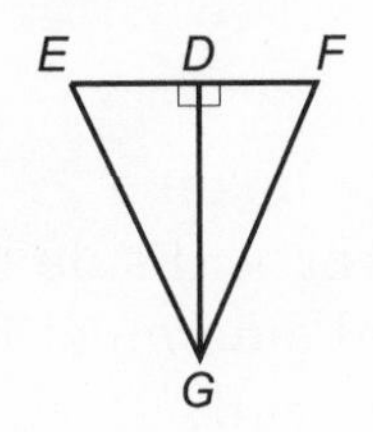

2. HA

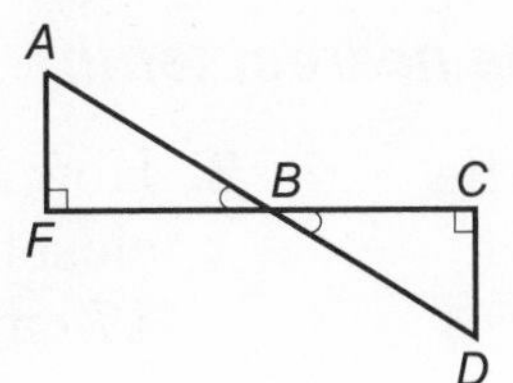

3. LL

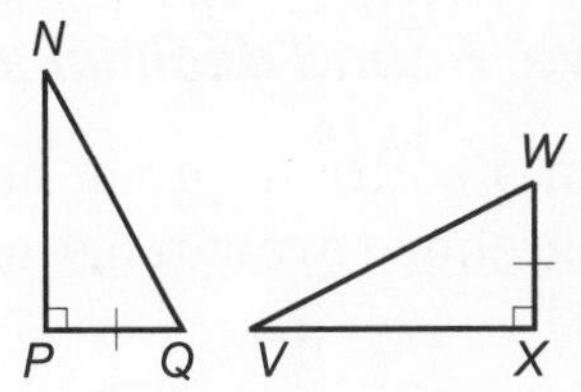

4. LA

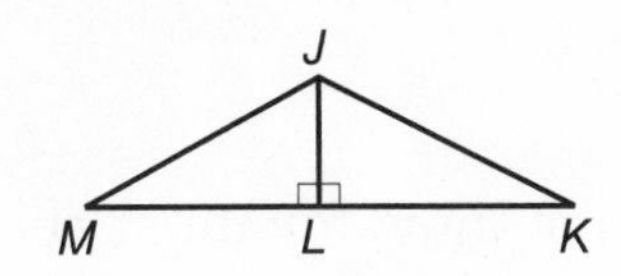

5. HA

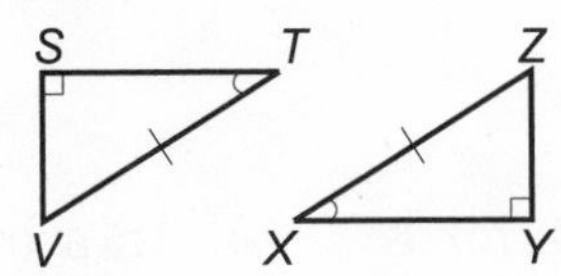

6. LA

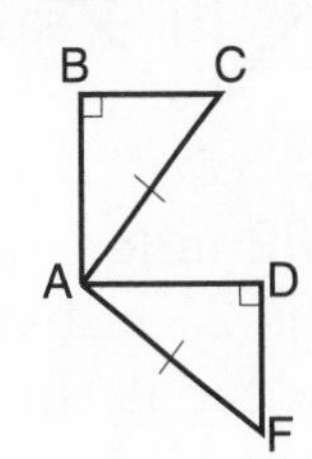

The Pythagorean Theorem

In a right triangle, the square of the hypotenuse, c, is equal to the sum of the squares of the lengths of the other two sides, a and b.

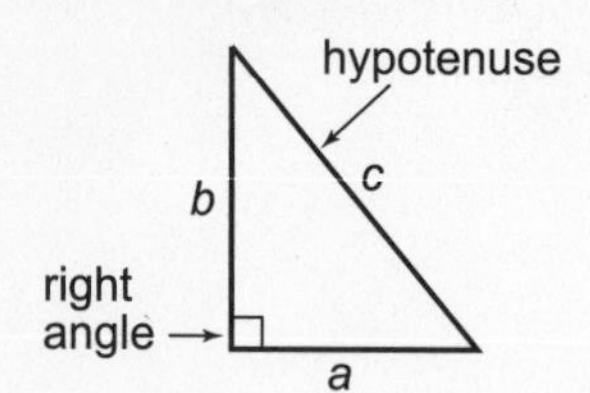

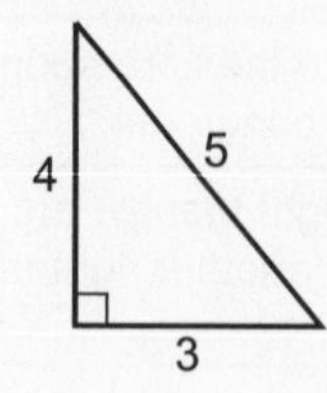

$$c^2 = a^2 + b^2$$
$$5^2 = 3^2 + 4^2$$
$$5^2 = 9 + 16$$
$$25 = 25$$

Example: How long must a ladder be to reach a window 13 feet above ground? For the sake of stability, the ladder must be placed 5 feet away from the base of the wall.

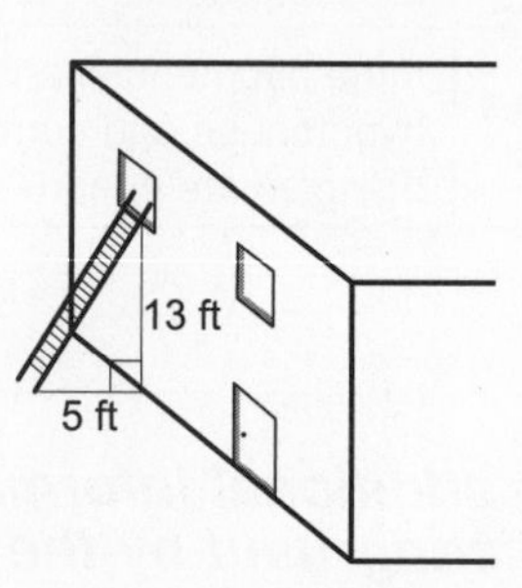

$$c^2 = (13)^2 + (5)^2$$
$$c^2 = 169 + 25$$
$$c^2 = 194$$
$$c^2 = \sqrt{194} \text{ or about } 13.9 \text{ ft}$$

Solve. Round decimal answers to the nearest tenth.

1. In a softball game, how far must the catcher throw to second base?

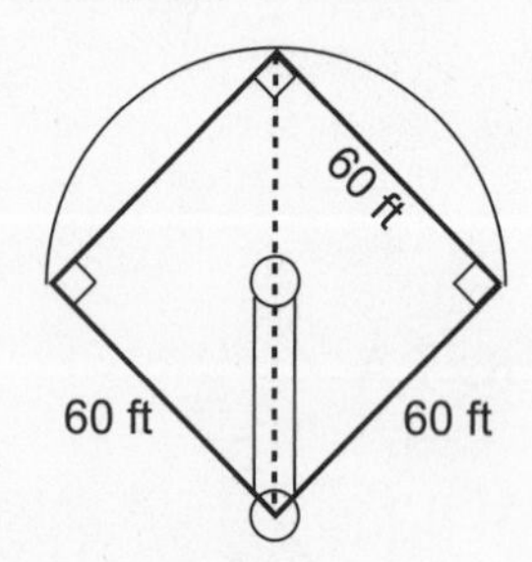

2. How long must the brace be on a closet rod holder if the vertical side is 17 cm and the horizontal side must be attached 30 cm from the wall?

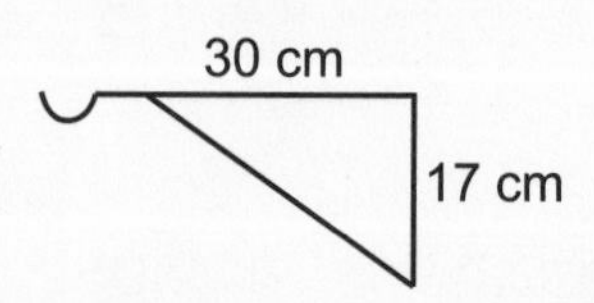

3. If Briny is 32 miles due east of Oxford and Myers is 21 miles due south of Oxford, how far is the shortest distance from Myers to Briny?

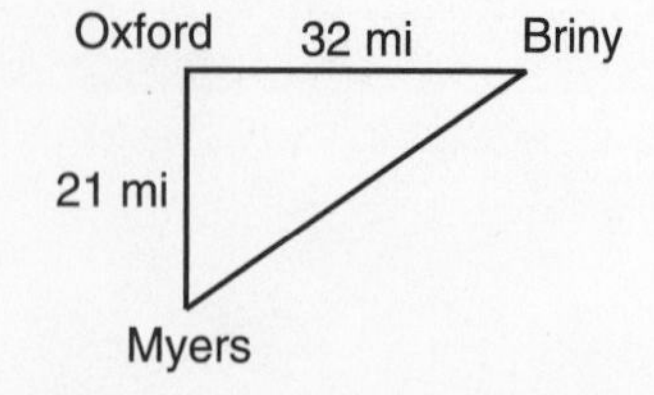

4. In a baseball game, how far must the shortstop (halfway between second base and third base) throw to make an out at first base?

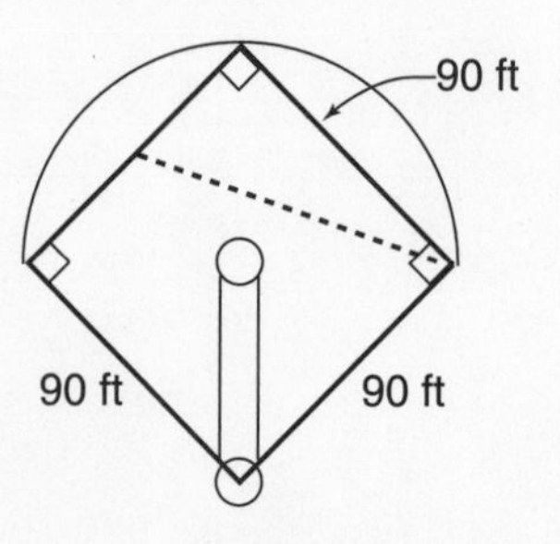

NAME ______________________ DATE __________ PERIOD ______

6-7 Study Guide

Student Edition
Pages 262–267

Distance on the Coordinate Plane

You can use the Pythagorean Theorem to find the distance between two points on the coordinate plane.

Example: Use the Distance Formula to find the distance between $R(2, -3)$ and $S(5, 4)$. Round to the nearest tenth if necessary.

Replace (x_1, y_1) with $(2, -3)$ and (x_2, y_2) with $(5, 4)$.

$$d = \sqrt{(x_2 - x_1)^2 + (y_2 - y_1)^2}$$
$$RS = \sqrt{(5 - 2)^2 + [4 - (-3)]^2}$$
$$RS = \sqrt{3^2 + (4 + 3)^2}$$
$$RS = \sqrt{3^2 + 7^2}$$
$$RS = \sqrt{9 + 49}$$
$$RS = \sqrt{58}$$
$$RS \approx 7.6$$

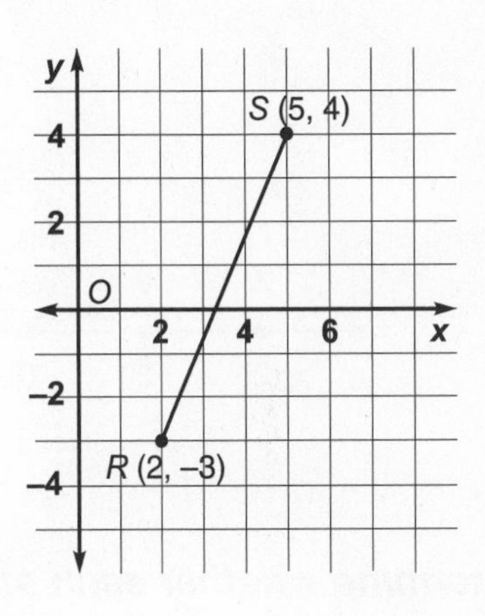

The distance between $R(2, -3)$ and $S(5, 4)$ is about 7.6 units.

Find the distance between each pair of points. Round to the nearest tenth, if necessary.

1.

2.

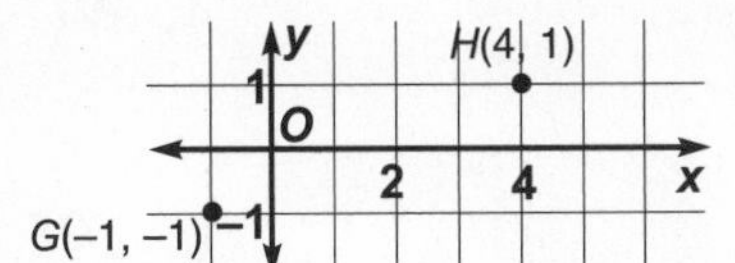

3.

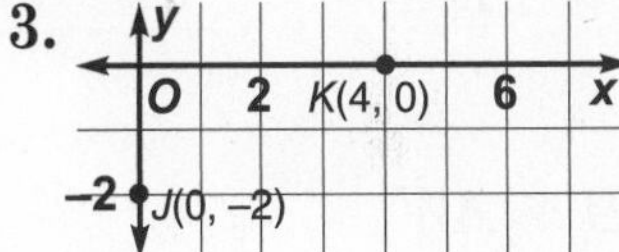

Graph each pair of ordered pairs. Then find the distance between the points. Round to the nearest tenth, if necessary.

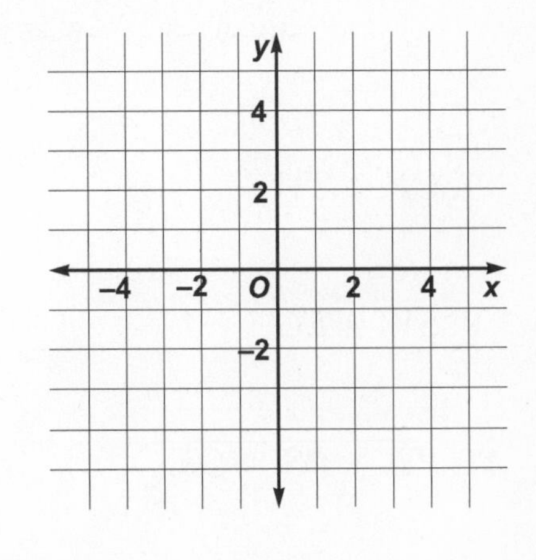

4. $A(4, 5); B(0, 2)$

5. $X(0, -4); Y(-3, 0)$

6. $M(3, 1); N(1, -4)$

7. $U(-1, 1); V(-4, 4)$

7-1 NAME ______________________ DATE __________ PERIOD ________

Study Guide

Student Edition
Pages 276–281

Segments, Angles, and Inequalities

Because measures are real numbers, you can use inequalities to compare two measures.

Symbol	Meaning
$\neq$	not equal to
$<$	less than
$>$	greater than
$\leq$	less than or equal to
$\geq$	greater than or equal to
$\nleq$	Not less than or equal to
$\ngeq$	not greater than or equal to

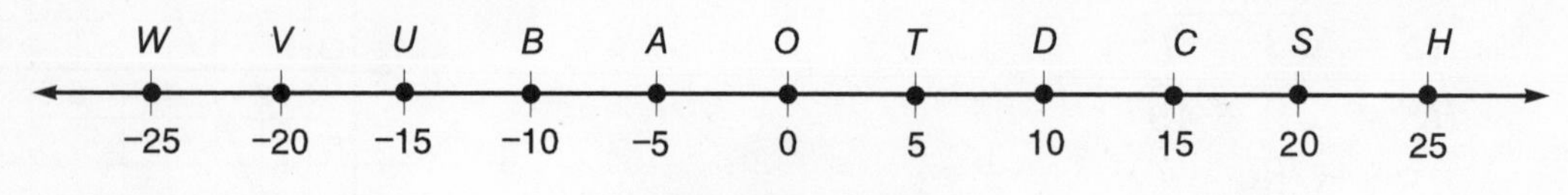

Exercises 1–6

Determine whether each statement is true or false.

1. $HC < BU$
2. $BC \neq HC$
3. $OH = BS$
4. $WU \leq AT$
5. $DB \ngtr AW$
6. $DH \ngeq BT$

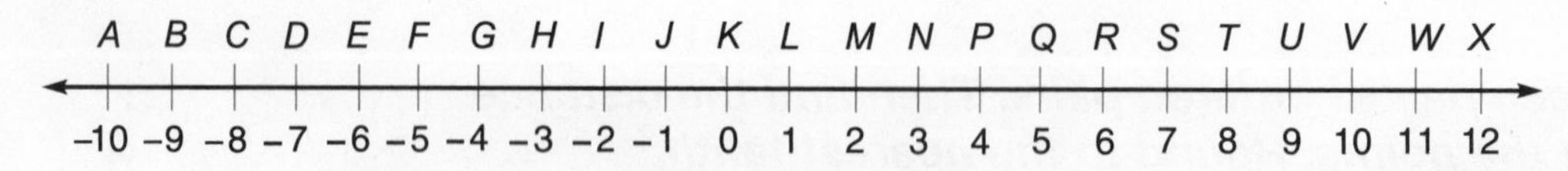

Exercises 7–12

7. $CF < TV$
8. $EP = JU$
9. $SW \neq EG$
10. $GI + IP > MV$
11. $PW - PS = GK$
12. $FT - KR = JS$

NAME ______________________________ DATE __________ PERIOD ______

Study Guide

Exterior Angle Theorem

An **exterior angle** of a triangle is an angle that forms a linear pair with one of the angles of the triangle. **Remote interior angles** of a triangle are the two angles that do *not* form a linear pair with the exterior angle. The measure of an exterior angle is equal to the sum of the measures of its two remote interior angles.

Example: $\angle 3$ is an exterior angle of $\angle RST$. $\angle 1$ and $\angle 4$ are remote interior angles with respect to $\angle 3$.

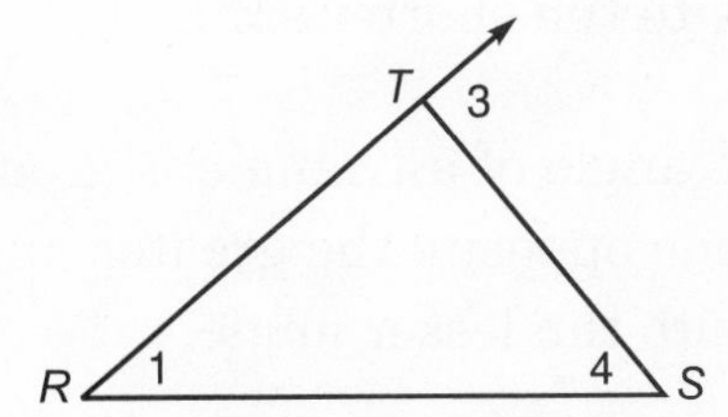

Use the figure at the right to fill in the blanks.

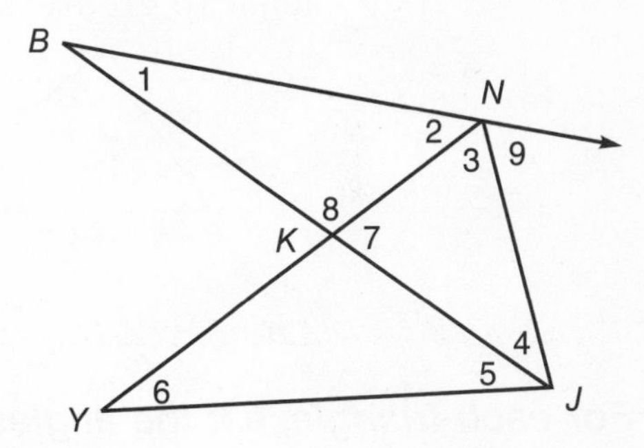

1. $\angle$________ is an exterior angle of $\triangle KYJ$.
2. $\angle$________ and $\angle$________ are remote interior angles with respect to exterior angle 8 in $\triangle KNJ$.
3. $\angle 8$ is an exterior angle of $\triangle$________.
4. $\angle 1$ and $\angle 4$ are remote interior angles with respect to exterior angle 9 in $\triangle$________.

Find the value of x in each triangle.

5.

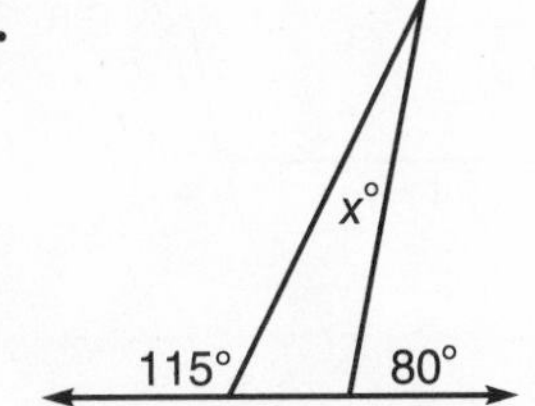

6.

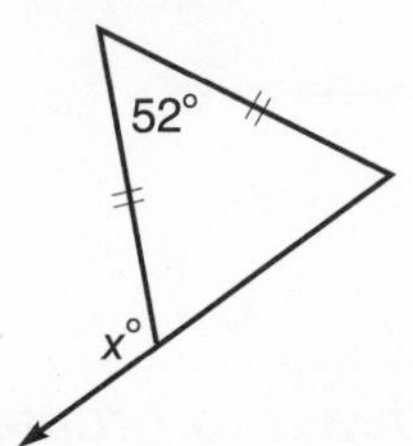

7.

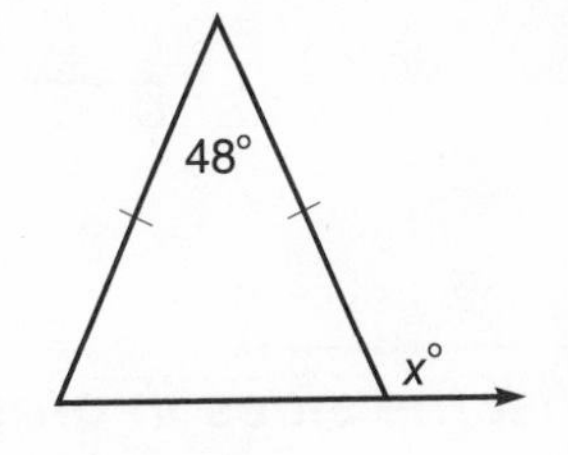

8.

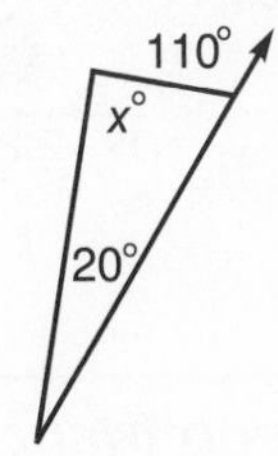

Replace each ● with <, >, or = to make a true sentence.

9. $m\angle 1$ ● $m\angle 6$

10. $m\angle 2$ ● $m\angle 1$

11. $m\angle 6$ ● $m\angle 3$

12. $m\angle 4$ ● $m\angle 6$

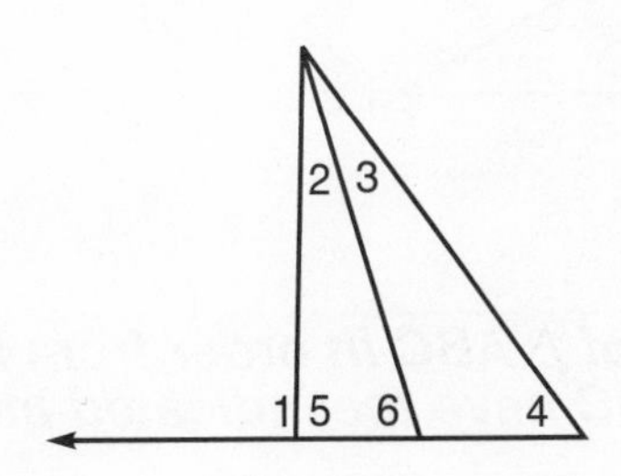

NAME ______________________ DATE __________ PERIOD ______

7-3 Study Guide

Student Edition
Pages 290–295

Inequalities Within a Triangle

Two theorems are very useful for determining relationships between sides and angles of triangles.

- If one side of a triangle is longer than another side, then the angle opposite the longer side is greater than the angle opposite the shorter side.
- If one angle of a triangle is greater than another angle, then the side opposite the greater angle is longer than the side opposite the lesser angle.

Examples: **1** List the angles in order from least to greatest measure.

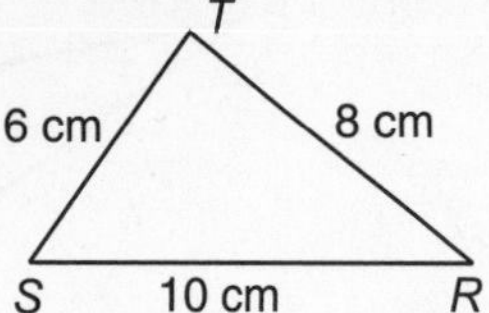

$\angle R, \angle S, \angle T$

2 List the sides in order from least to greatest measure.

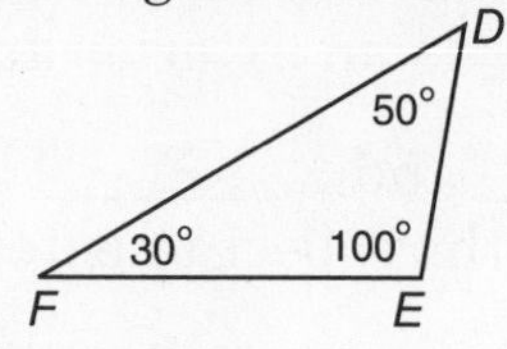

$\overline{DE}, \overline{FE}, \overline{FD}$

For each triangle, list the angles in order from least to greatest measure.

1.

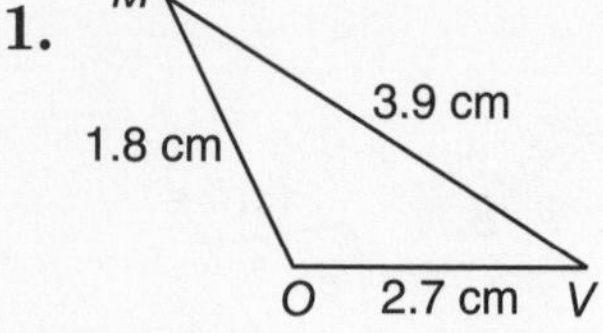

2.

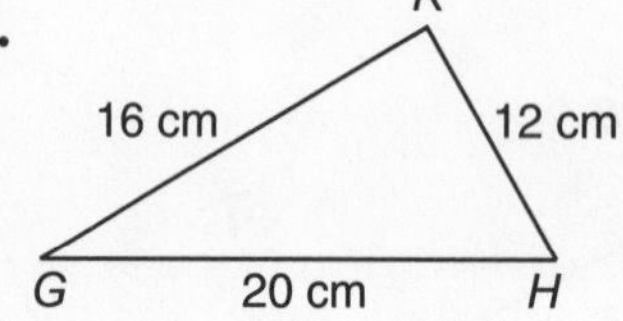

3.

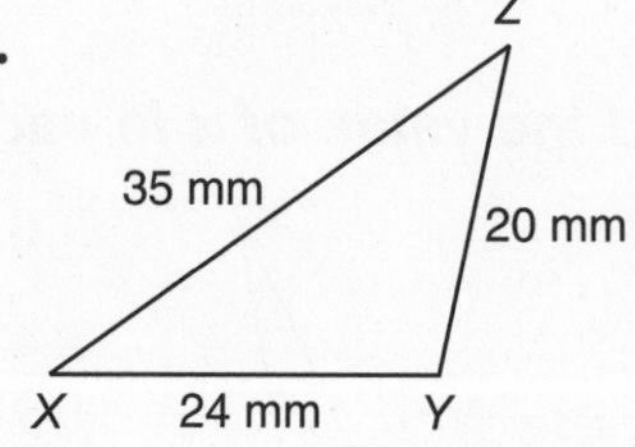

For each triangle, list the sides in order from least to greatest measure.

4.

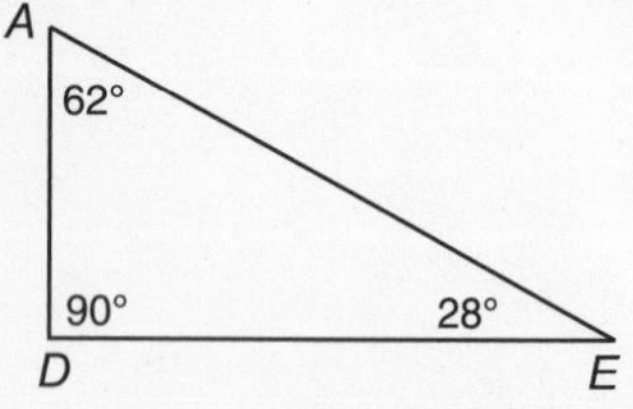

5.

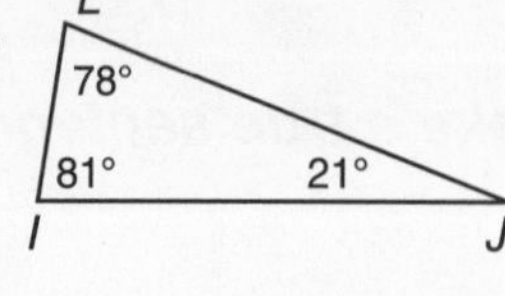

6.

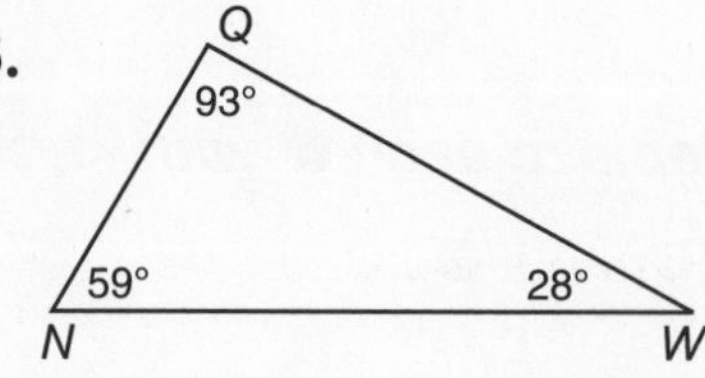

List the sides of $\triangle ABC$ in order from longest to shortest if the angles of $\triangle ABC$ have the indicated measures.

7. $m\angle A = 5x + 2$, $m\angle B = 6x - 10$, $m\angle C = x + 20$

8. $m\angle A = 10x$, $m\angle B = 5x - 17$, $m\angle C = 7x - 1$

NAME ______________________ DATE __________ PERIOD ______

Study Guide

Student Edition
Pages 296–301

Triangle Inequality Theorem

If you take three straws that are 8 inches, 4 inches, and 3 inches in length, can you use these three straws to form a triangle? Without actually trying it, you might think it is possible to form a triangle with the straws. If you try it, however, you will notice that the two smaller straws are too short. This example illustrates the following theorem.

Triangle Inequality Theorem	The sum of the measures of any two sides of a triangle is greater than the measure of the third side.

Example: If the lengths of two sides of a triangle are 7 centimeters and 11 centimeters, between what two numbers must the measure of the third side fall?

Let x = the length of the third side.

By the Triangle Inequality Theorem, each of these inequalities must be true.

$x + 7 > 11$ | $x + 11 > 7$ | $11 + 7 > x$
$x > 4$ | $x > -4$ | $18 > x$

Therefore, x must be between 4 centimeters and 18 centimeters.

***Determine whether it is possible to draw a triangle with sides of the given measures. Write* yes *or* no.**

1. 15, 12, 9 **2.** 23, 16, 7 **3.** 20, 10, 9

4. 8.5, 6.5, 13.5 **5.** 47, 28, 70 **6.** 28, 41, 13

The measures of two sides of a triangle are given. Between what two numbers must the measure of the third side fall?

7. 9 and 15 **8.** 11 and 20 **9.** 23 and 14

10. Suppose you have three different positive numbers arranged in order from greatest to least. Which sum is it most crucial to test to see if the numbers could be the lengths of the sides of a triangle?

NAME ______________________ DATE __________ PERIOD ______

8-1 Study Guide

Student Edition
Pages 310–315

Quadrilaterals

A **quadrilateral** is a closed geometric figure with four sides and four vertices. Any two sides, vertices, or angles of a quadrilateral are said to be either **consecutive** or **opposite**. A segment joining any two nonconsecutive vertices in a quadrilateral is called a **diagonal**.

Refer to quadrilateral WXYZ for Exercises 1–5.

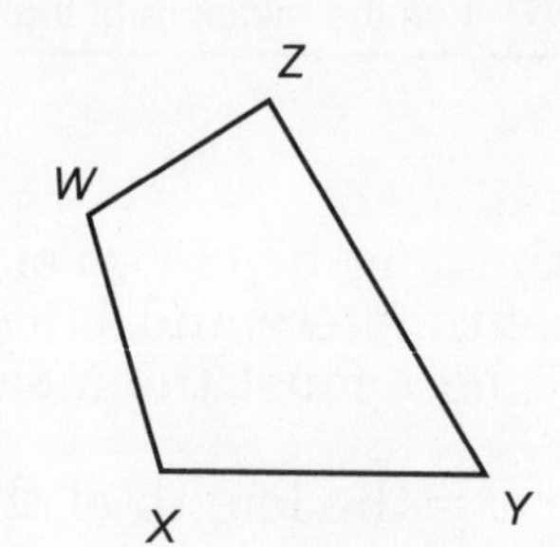

1. Name all pairs of opposite sides.

2. Name all pairs of consecutive angles.

3. Name the diagonals.

4. Name all pairs of consecutive vertices.

5. Name all pairs of opposite angles.

Find the missing measure(s) in each figure.

6.

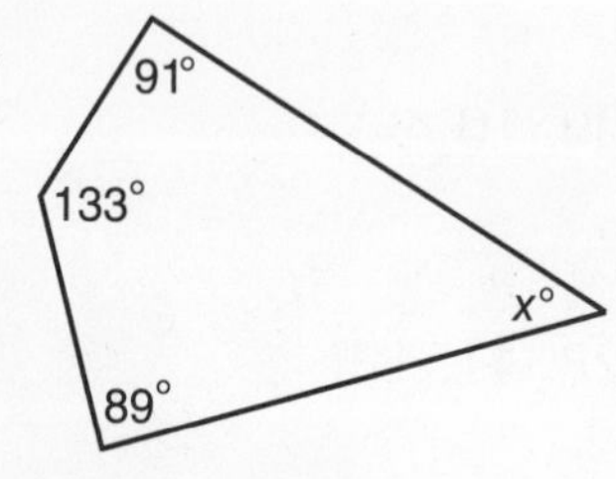

7.

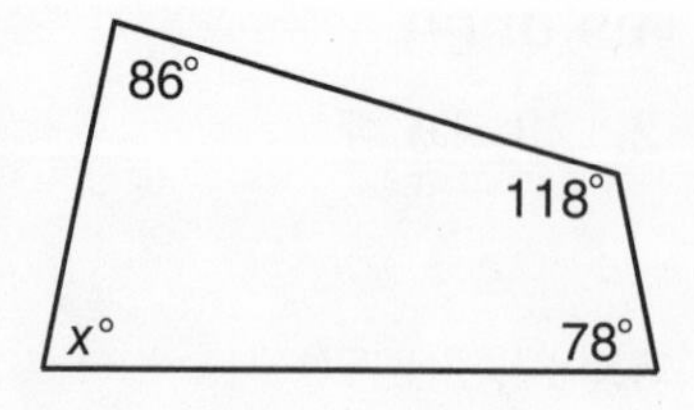

8.

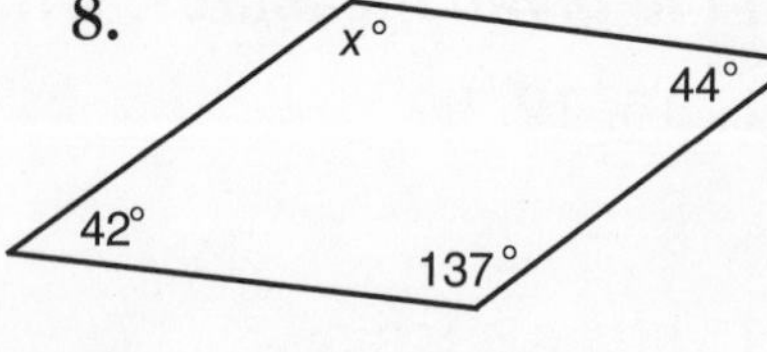

9.

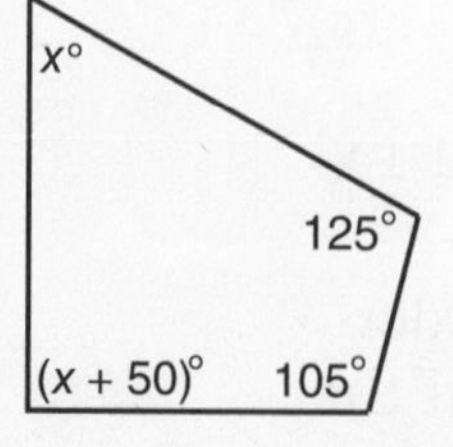

10.

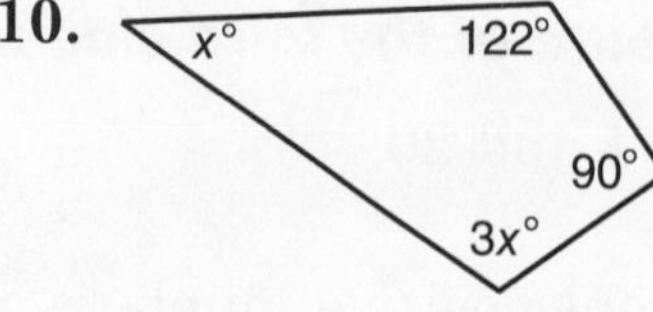

11.

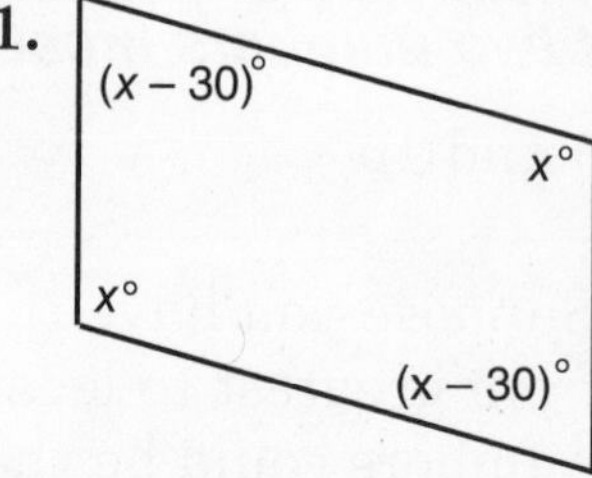

NAME ______________________ DATE __________ PERIOD ______

Study Guide

Student Edition
Pages 316–321

Parallelograms

A special kind of quadrilateral in which both pairs of opposite sides are parallel is called a **parallelogram**.

The following theorems all concern parallelograms.

- Opposite sides of a parallelogram are congruent.
- Opposite angles of a parallelogram are congruent.
- Consecutive angles of a parallelogram are supplementary.
- The diagonals of a parallelogram bisect each other.

Example: If the quadrilateral in the figure is a parallelogram, find the values of x, y, and z.

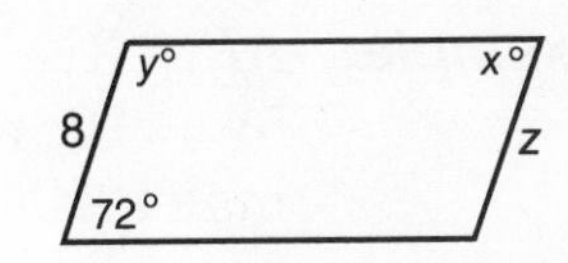

Since opposite angles of a parallelogram are congruent, $x = 72$.

Since consecutive angles of a parallelogram are supplementary, $y + 72 = 180$. Therefore, $y = 108$.

Since opposite sides of a parallelogram are congruent, $z = 8$.

If each quadrilateral is a parallelogram, find the values of x, y, and z.

1.

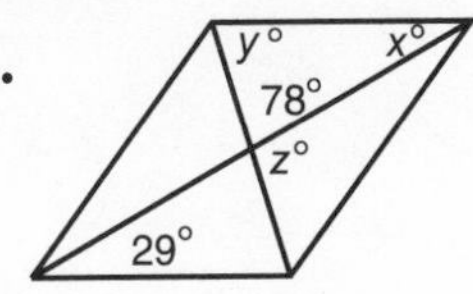

2.

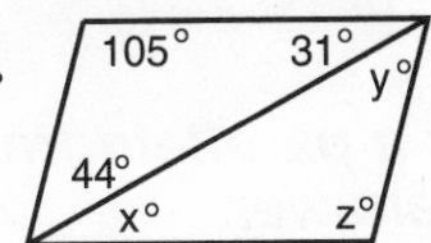

3.

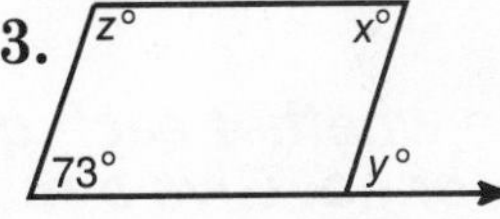

4. In parallelogram $ABCD$, $m\angle A = 3x$ and $m\angle B = 4x + 40$. Find the measure of angles A, B, C, and D.

5. In parallelogram $RSTV$, diagonals $\overline{RT}$ and $\overline{VS}$ intersect at Q. If $RQ = 5x + 1$ and $QT = 3x + 15$, find QT.

Explain why it is impossible for each figure to be a parallelogram.

6.

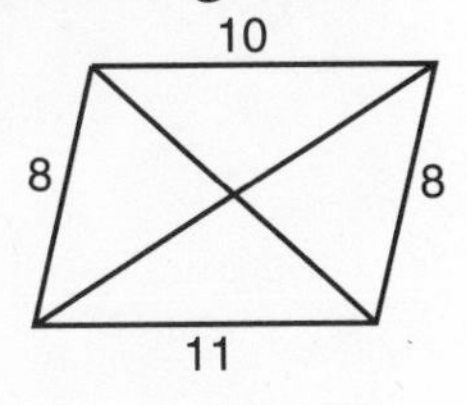

7. 130° 55° 50° 125°

8-3

NAME ______________________ DATE __________ PERIOD ______

Study Guide

Student Edition
Pages 322–326

Tests for Parallelograms

You can show that a quadrilateral is a parallelogram if you can show that one of the following is true.

1. Both pairs of opposite sides are parallel.
2. Both pairs of opposite sides are congruent.
3. Diagonals bisect each other.
4. Both pairs of opposite angles are congruent.
5. A pair of opposite sides is both parallel and congruent.

Example: $AP = 3x - 4$, $AC = 46$, $PB = 3y$, and $DP = 5y - 12$. Find the values of x and y that would make $ABCD$ a parallelogram.

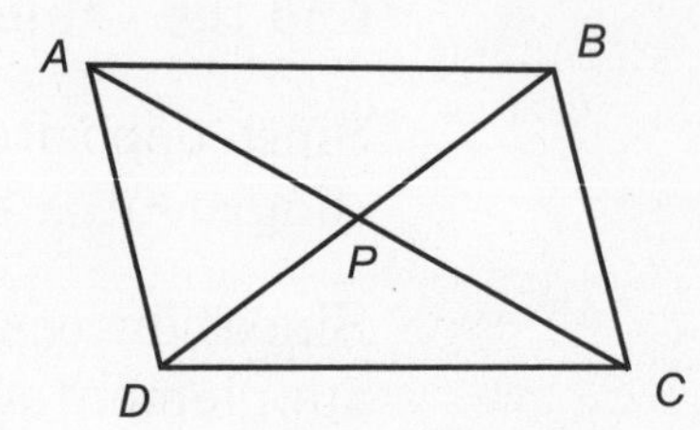

For the diagonals to bisect each other, $2(3x - 4) = 46$ and $3y = 5y - 12$. Solve for each variable.

$2(3x - 4) = 46$		$3y = 5y - 12$	
$6x - 8 = 46$	*Dist. Prop.*	$-2y = -12$	*Subtract 5y.*
$6x = 54$	*Add 8.*	$y = 6$	*Divide by −2*
$x = 9$	*Divide by 6.*		

So, $x = 9$ and $y = 6$.

Determine whether each quadrilateral is a parallelogram. Write yes or no. Give a reason for your answer.

1. 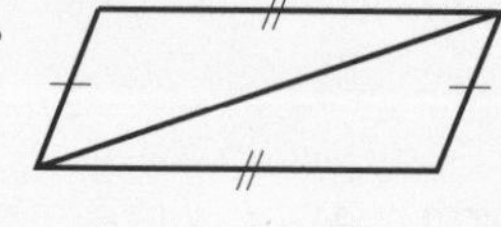

2. 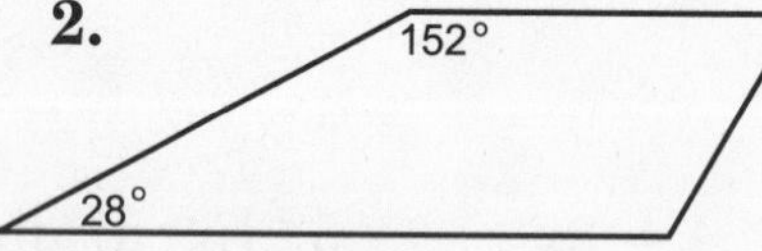

3. 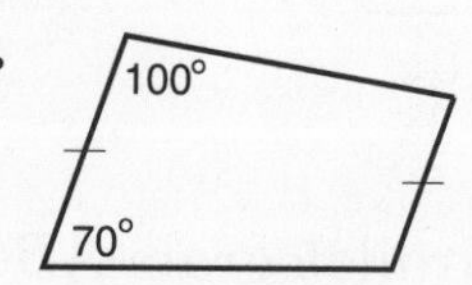

Find the values of x and y that ensure each quadrilateral is a parallelogram.

4. 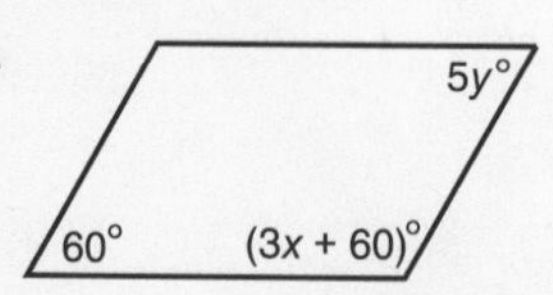

5. 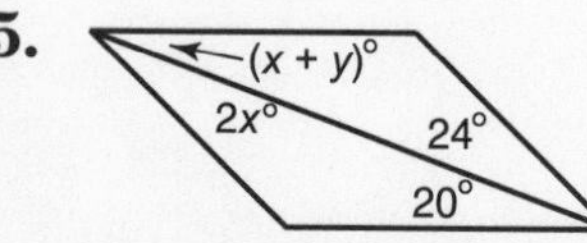

NAME ______________________ DATE __________ PERIOD _____

8-4 Study Guide

Student Edition
Pages 327–332

Rectangles, Rhombi, and Squares

A **rectangle** is a quadrilateral with four right angles. A **rhombus** is a quadrilateral with four congruent sides. A **square** is a quadrilateral with four right angles and four congruent sides. A square is both a rectangle and a rhombus. Rectangles, rhombi, and squares are all examples of parallelograms.

Rectangles	Rhombi
• Opposite sides are congruent. • Opposite angles are congruent. • Consecutive angles are supplementary. • Diagonals bisect each other. • All four angles are right angles. • Diagonals are congruent.	• Diagonals are perpendicular. • Each diagonal bisects a pair of opposite angles.

Determine whether each statement is* always, sometimes, *or* never *true.

1. The diagonals of a rectangle are perpendicular.

2. Consecutive sides of a rhombus are congruent.

3. A rectangle has at least one right angle.

4. The diagonals of a parallelogram are congruent.

5. A diagonal of a square bisects opposite angles.

***Use rhombus DLMP to determine whether each statement is* true *or* false.**

6. $OM = 13$

7. $PL = 26$

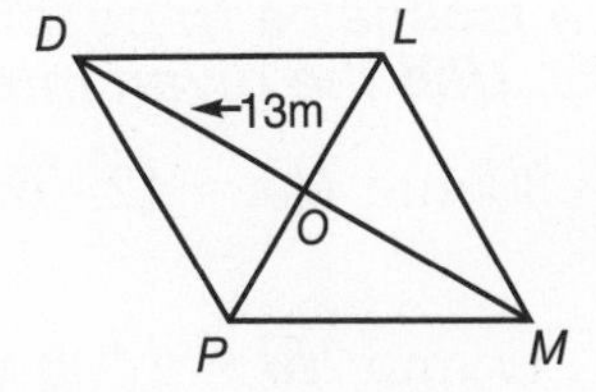

8. $\overline{MD} \cong \overline{PL}$

9. $m\angle DLO = m\angle LDO$

10. $\angle LDP \cong \angle LMP$

11. $m\angle DPM = m\angle PML$

NAME ______________________ DATE __________ PERIOD ______

Study Guide

Student Edition
Pages 333–339

Trapezoids

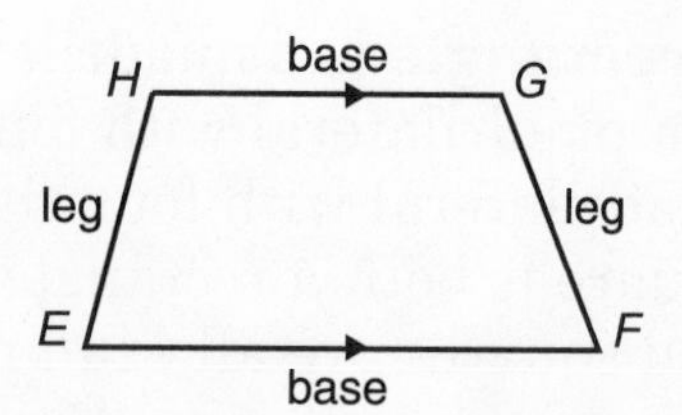

A **trapezoid** is a quadrilateral with exactly one pair of parallel sides. The parallel sides are called **bases**, and the nonparallel sides are called **legs**. In trapezoid *EFGH*, $\angle E$ and $\angle F$ are called **base angles**. $\angle H$ and $\angle G$ form the other pair of base angles.

A trapezoid is an **isosceles trapezoid** if its legs are congruent.

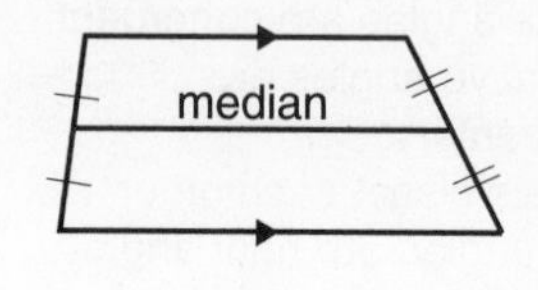

The **median** of a trapezoid is the segment that joins the midpoints of the legs.

The following theorems all concern trapezoids.

- Both pairs of base angles of an isosceles trapezoid are congruent.
- The diagonals of an isosceles trapezoid are congruent.
- The median of a trapezoid is parallel to the bases, and its measure is one-half the sum of the measures of the bases.

Example: Given trapezoid *RSTV* with median $\overline{MN}$, find the value of x.

$$MN = \frac{1}{2}(VT + RS)$$
$$15 = \frac{1}{2}(6x - 3 + 8x + 5)$$
$$15 = \frac{1}{2}(14x + 2)$$
$$15 = 7x + 1$$
$$14 = 7x$$
$$2 = x$$

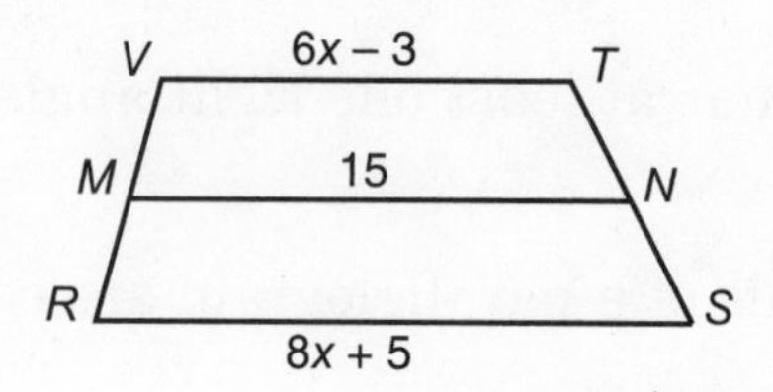

HJKL is an isosceles trapezoid with bases $\overline{HJ}$ and $\overline{LK}$, and median RS. Use the given information to solve each problem.

1. If $LK = 30$ and $HJ = 42$, find RS.

2. If $RS = 17$ and $HJ = 14$, find LK.

3. If $RS = x + 5$ and $HJ + LK = 4x + 6$, find RS.

4. If $m\angle LRS = 66$, find $m\angle KSR$.

NAME ______________________ DATE __________ PERIOD ______

Study Guide

Student Edition
Pages 350–355

Using Ratios and Proportions

A **ratio** is a comparison of two quantities. The ratio of a to b can be expressed as $\frac{a}{b}$, where b is not 0. The ratio can also be written $a{:}b$.

An equation stating that two ratios are equal is a **proportion**. Therefore, $\frac{a}{b} = \frac{c}{d}$ is a proportion for any numbers a and c and any nonzero numbers b and d. In any true proportion, the cross products are equal. So, $\frac{a}{b} = \frac{c}{d}$ if and only if $ad = bc$.

Example: Solve $\frac{11}{16} = \frac{44}{x}$ by using cross products.

$$\frac{11}{16} = \frac{44}{x}$$
$$11x = 16 \cdot 44$$
$$11x = 704$$
$$x = 64$$

For Exercises 1–4, use the table to find the ratios. Express each ratio as a decimal rounded to three places.

Teams	Wins	Losses
Hawks	16	13
Tigers	15	14
Mustangs	12	16

1. games won to games lost for Hawks

2. games won by the Hawks to games won by Tigers

3. games won to games played for Tigers

4. games won to games played for Mustangs

Solve each proportion by using cross products.

5. $\frac{9}{28} = \frac{x}{84}$

6. $\frac{3}{18} = \frac{4x}{7}$

7. $\frac{x+5}{7} = \frac{x+3}{5}$

Use a proportion to solve each problem.

8. If two cassettes cost $14.50, how much will 15 cassettes cost?

9. If a 6-foot post casts a shadow that is 8 feet long, how tall is an antenna that casts a 60-foot shadow at the same time?

NAME ______________________ DATE __________ PERIOD ______

Study Guide

Similar Polygons

Two polygons are **similar** if and only if their corresponding angles are congruent and the measures of their corresponding sides are proportional.

The symbol ~ means *is similar to.*

The ratio of the lengths of two corresponding sides of two similar polygons is called the **scale factor**.

Example: Find x if $\triangle RST \sim \triangle XYZ$.

The corresponding sides are proportional, so we can write a proportion to find the value of x.

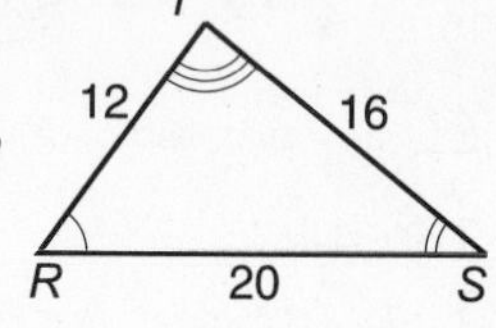

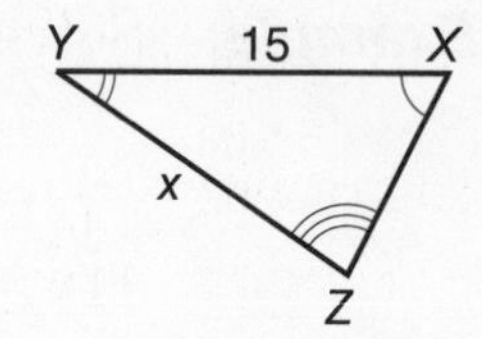

$$\frac{16}{x} = \frac{20}{15}$$
$$20x = 240$$
$$x = 12$$

If quadrilateral ABCD is similar to quadrilateral EFGH, find each of the following.

1. scale factor of $ABCD$ to $EFGH$

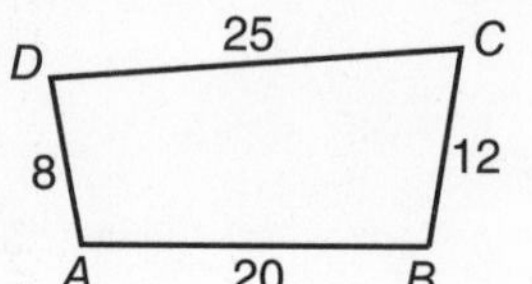

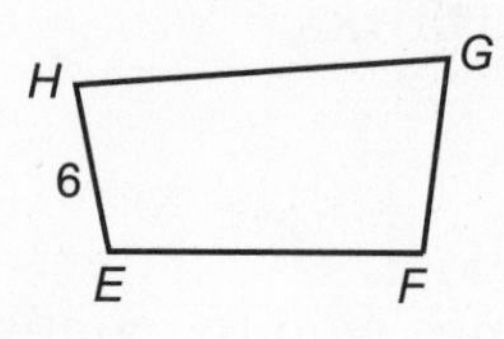

2. EF
3. FG
4. GH
5. perimeter of $ABCD$
6. perimeter of $EFGH$
7. ratio of perimeter of $ABCD$ to perimeter of $EFGH$

Each pair of polygons is similar. Find the values of x and y.

8.

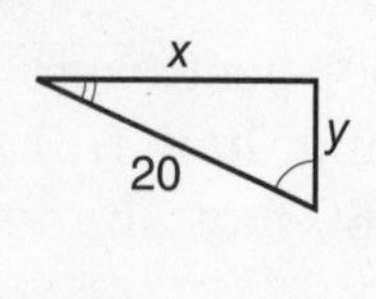

9.

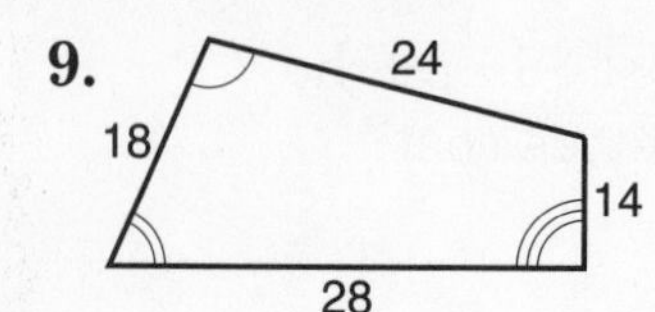

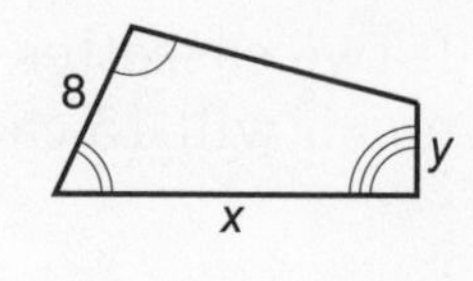

NAME ______________________ DATE ________ PERIOD ______

Study Guide

Student Edition
Pages 362–367

Similar Triangles

There are three ways to determine whether two triangles are similar.

- Show that two angles of one triangle are congruent to two angles of the other triangle. (AA Similarity)
- Show that the measures of the corresponding sides of the triangles are proportional. (SSS Similarity)
- Show that the measure of two sides of a triangle are proportional to the measures of the corresponding sides of the other triangle and that the included angles are congruent. (SAS Similarity)

Example: Determine whether the triangles are similar. Explain your answer.

Since $\frac{15}{12} = \frac{25}{20} = \frac{20}{16}$, the triangles are similar by SSS Similarity.

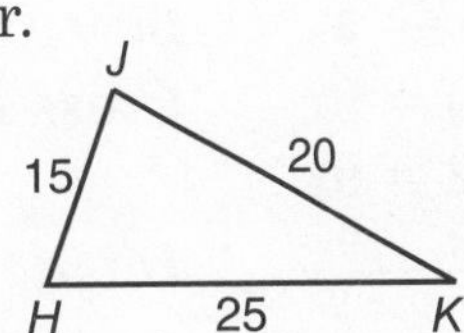

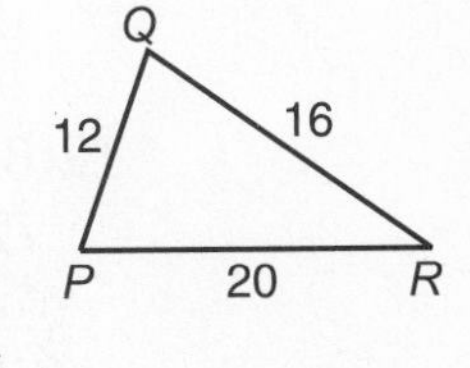

Determine whether each pair of triangles is similar. Give a reason for your answer.

1.

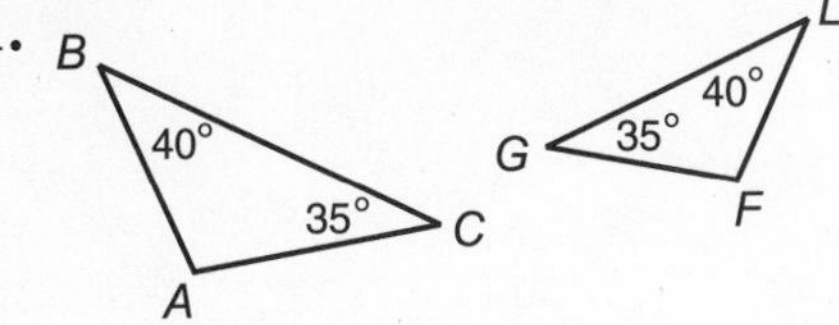

2.

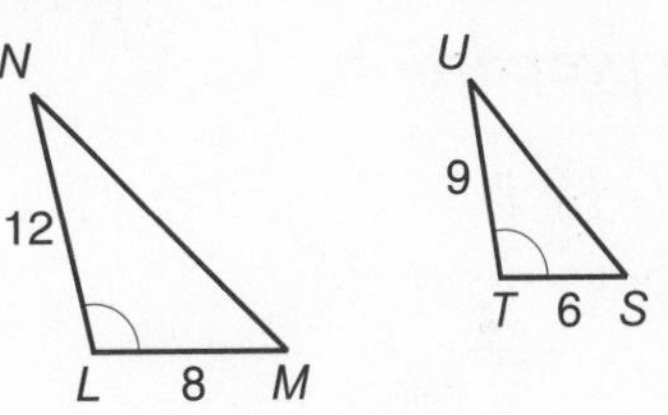

3.

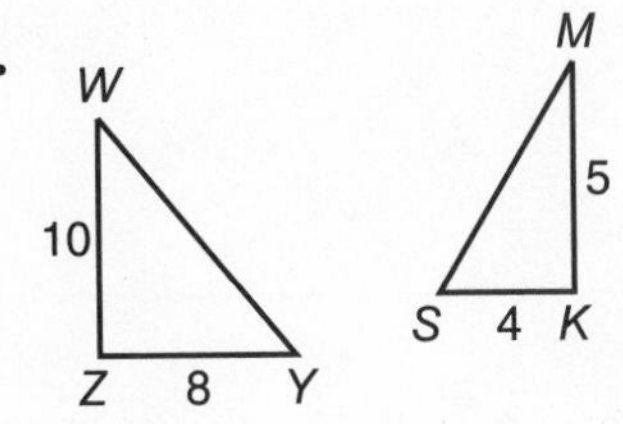

4.

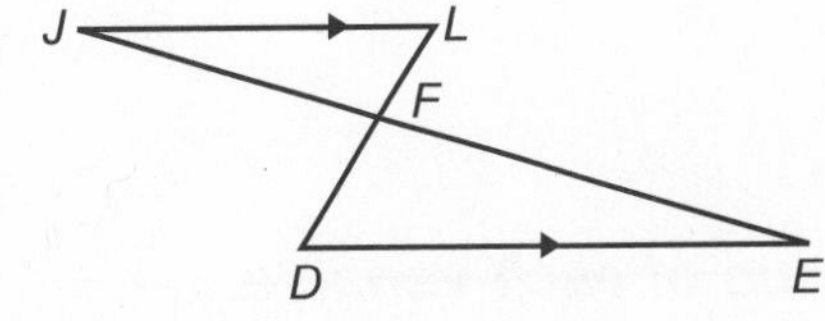

Identify the similar triangles in each figure. Explain why they are similar and find the missing measures.

5. If $\overline{MN} \parallel \overline{AB}$, find AB, BC, and BN.

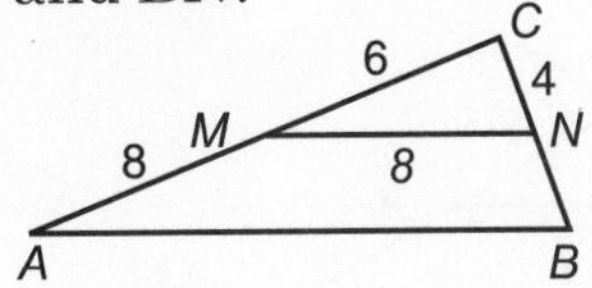

6. If $MNPQ$ is a parallelogram, find RN, RP, and SP.

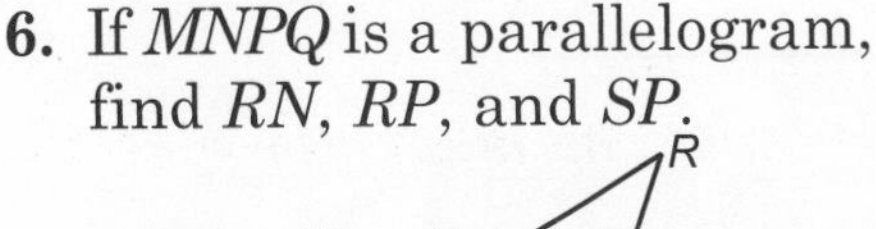

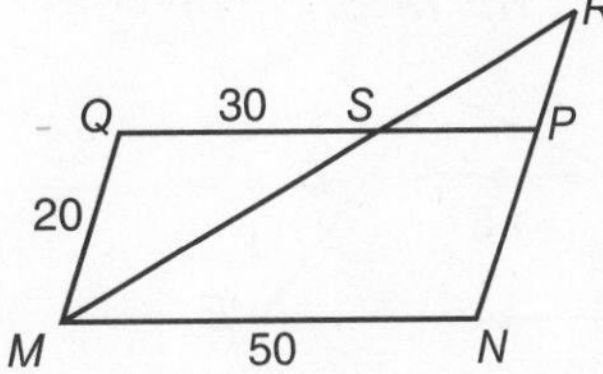

NAME ______________________ DATE __________ PERIOD _____

Study Guide

Student Edition
Pages 368-373

Proportional Parts and Triangles

If a line is parallel to one side of a triangle and intersects the other two sides, then:

- the triangle formed is similar to the original triangle, and
- the line separates the sides into segments of proportional lengths.

Example: In the figure, $\overline{AB} \parallel \overline{DE}$. Find the value of x.

$\frac{AD}{DC} = \frac{BE}{EC}$	*Theorem 9-5*
$\frac{12}{9} = \frac{x}{6}$	*AD = 12, BE = x, DC = 9, EC = 6*
$12(6) = 9x$	*Cross Products*
$72 = 9x$	
$\frac{72}{9} = \frac{9x}{9}$	*Divide each side by 9.*
$8 = x$	

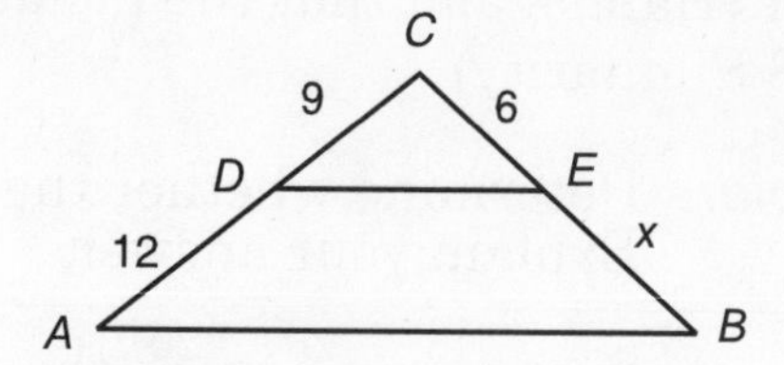

Complete each proportion.

1. $\frac{AB}{BM} = \frac{?}{CD}$

2. $\frac{AD}{AC} = \frac{AM}{?}$

3. $\frac{CB}{DM} = \frac{AC}{?}$

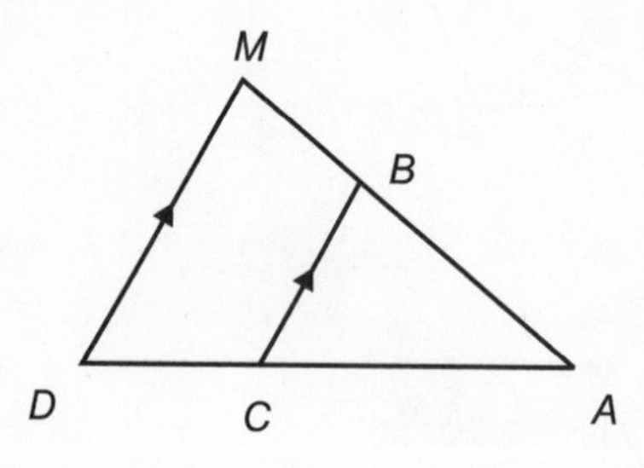

Find the value of each variable.

4.

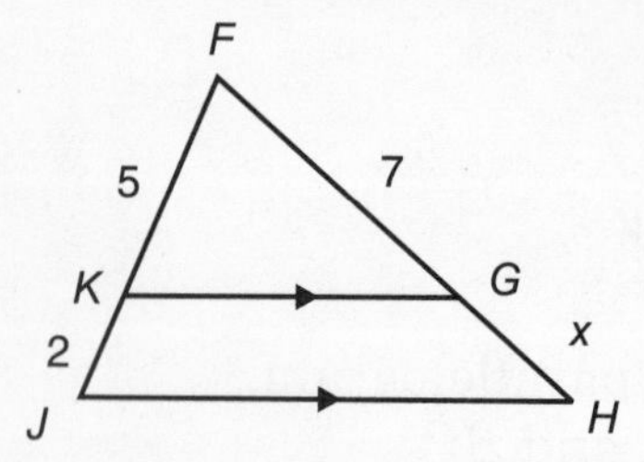

5.

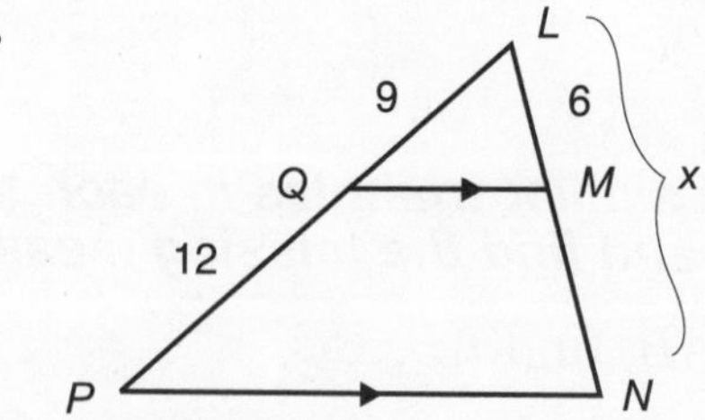

6.

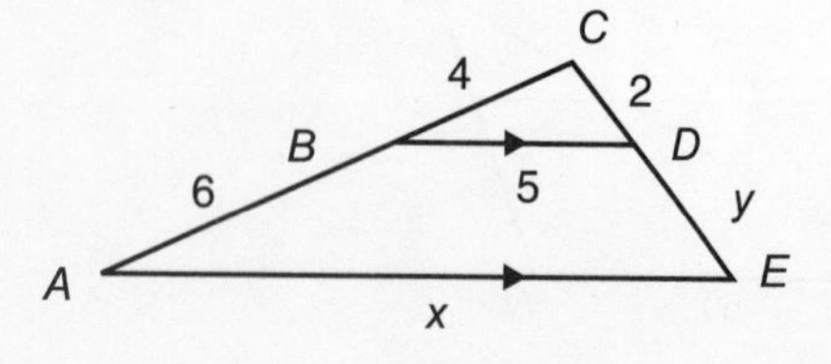

7.

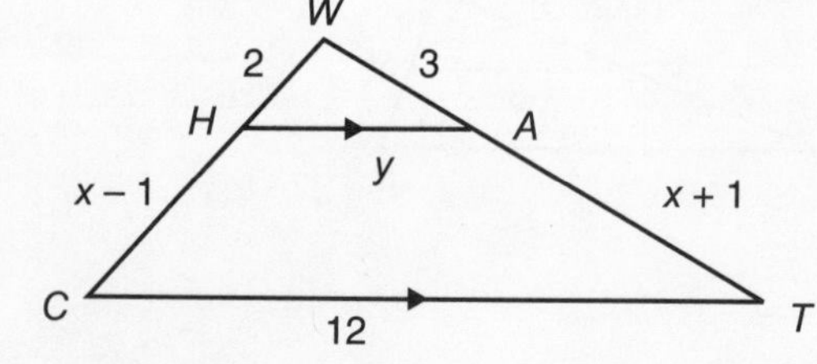

NAME ______________________ DATE __________ PERIOD ______

Study Guide

Student Edition
Pages 374–379

Triangles and Parallel Lines

The following theorems involve proportional parts of triangles.

- If a line is parallel to one side of a triangle and intersects the other two sides, then it separates these sides into segments of proportional lengths.
- A segment whose endpoints are the midpoints of two sides of a triangle is parallel to the third side of the triangle and its length is one-half the length of the third side.

Example: In $\triangle ABC$, $\overline{EF} \parallel \overline{CB}$, find the value of x.

$\overline{EF} \parallel \overline{CB}$ implies that $\frac{AF}{FB} = \frac{AE}{EC}$.

Rewrite the proportion and solve.

$$\frac{x + 20}{x + 8} = \frac{18}{10}$$
$$10x + 200 = 18x + 144$$
$$56 = 8x$$
$$7 = x$$

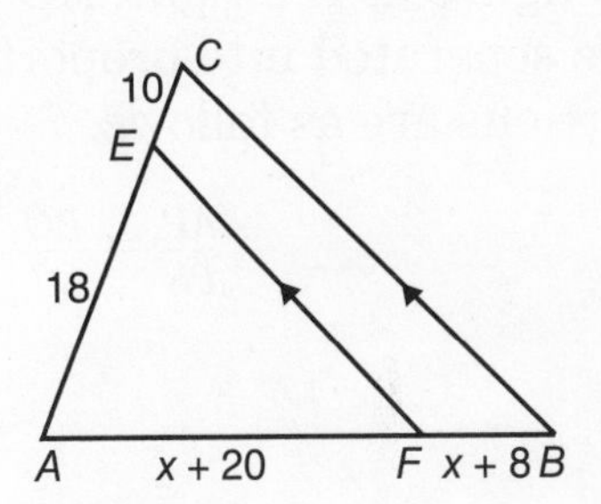

Find the value of x.

1. 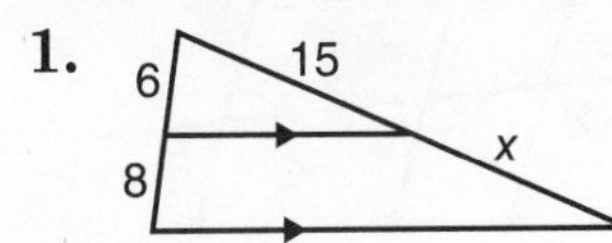

2. x + 6
30
x
9

In △ABC, find x so that $\overline{DE} \parallel \overline{CB}$.

3. $DC = 18$, $AD = 6$, $AE = 12$, $EB = x - 3$

4. $AC = 30$, $AD = 10$, $AE = 22$, $EB = x + 4$

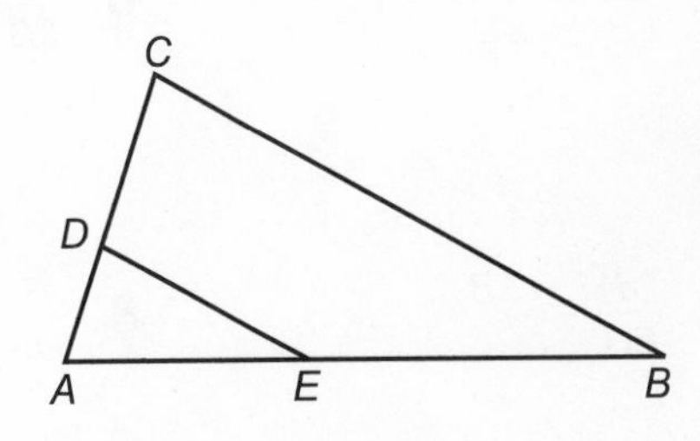

5. In $\triangle RST$, M is the midpoint of $\overline{RS}$, N is the midpoint of $\overline{ST}$, and P is the midpoint of $\overline{RT}$. Find the perimeter of $\triangle MNP$ if $RS = 28$, $ST = 34$, and $RT = 30$.

NAME ______________________ DATE ________ PERIOD _____

9-6 Study Guide

Student Edition
Pages 382–387

Proportional Parts and Parallel Lines

The following theorems involve parallel lines and proportional parts.

- If three or more parallel lines intersect two transversals, then they divide the transversals proportionally.
- If three or more parallel lines cut off congruent segments on one transversal, then they cut off congruent segments on every transversal.

In the figure at the right, $\overleftrightarrow{AB} \parallel \overleftrightarrow{PQ} \parallel \overleftrightarrow{ST}$. Transversals $\overleftrightarrow{AS}$ and $\overleftrightarrow{BT}$ are separated into proportional segments. Some of these proportions are as follows.

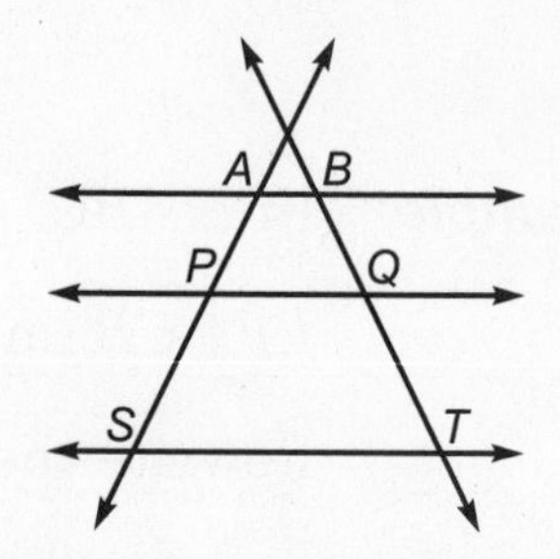

$$\frac{AP}{PS} = \frac{BQ}{QT}, \frac{AS}{BT} = \frac{PS}{QT}, \text{ and } \frac{SA}{PA} = \frac{TB}{QB}$$

Complete each proportion.

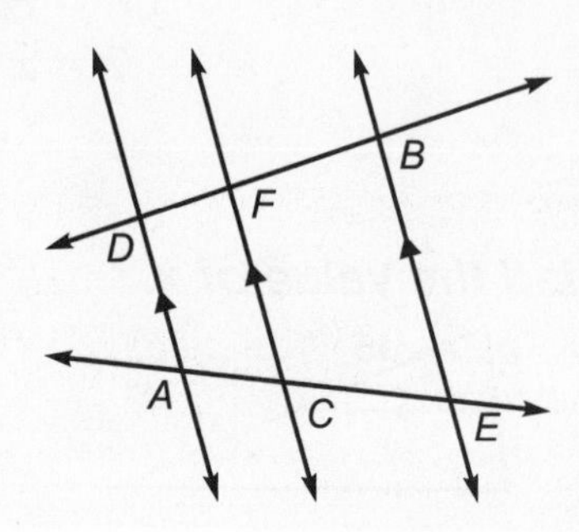

1. $\frac{DF}{FB} = \frac{?}{CE}$

2. $\frac{FB}{?} = \frac{DF}{AC}$

3. $\frac{AC}{?} = \frac{CE}{FB}$

4. $\frac{CE}{AC} = \frac{FB}{?}$

***Determine whether each statement is* true *or* false.**

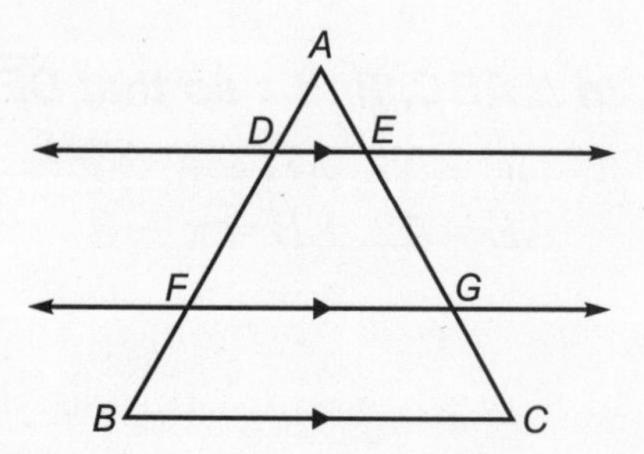

5. $\frac{AF}{FB} = \frac{AE}{EC}$

6. $\frac{DF}{FB} = \frac{EG}{GC}$

7. $\frac{DF}{EG} = \frac{FB}{GC}$

8. $\frac{AD}{DB} = \frac{AE}{EC}$

9. $\frac{DE}{FG} = \frac{AE}{EC}$

10. $\frac{AG}{AC} = \frac{FG}{BC}$

NAME ______________________________ DATE __________ PERIOD ______

Study Guide

Perimeters and Similarity

If two triangles are similar, then the measures of the corresponding perimeters are proportional to the measures of the corresponding sides.

Example: Determine the scale factor of $\triangle ABC$ to $\triangle DEF$.

$\frac{AB}{DE} = \frac{16}{4}$ or 4 $\quad \frac{BC}{EF} = \frac{20}{5}$ or 4

$\frac{CD}{FD} = \frac{24}{6}$ or 4

The scale factor is 4.

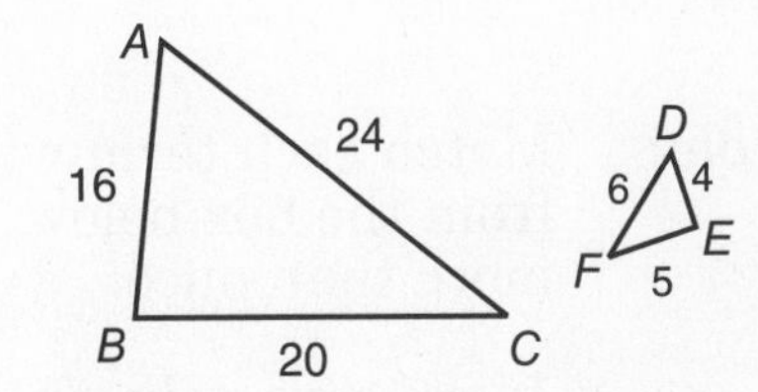

In the figure, $\triangle RST \sim \triangle XYZ$.

1. Find the scale factor of $\triangle XYZ$ to $\triangle RST$.

2. Find the perimeter of $\triangle RST$.

3. Find the perimeter of $\triangle XYZ$.

4. What is the ratio of the perimeters of $\triangle RST$ to $\triangle XYZ$?

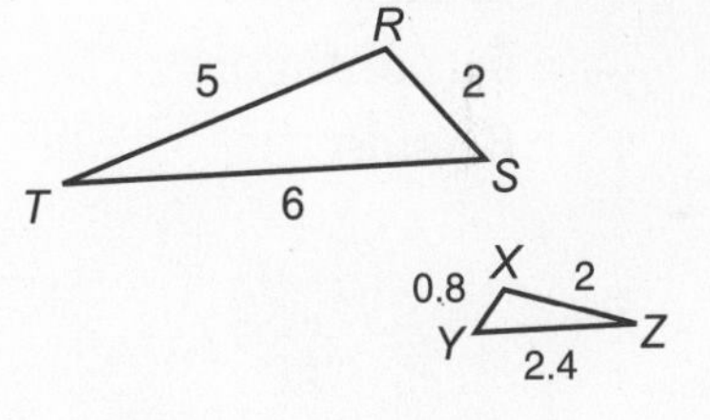

In the figure, $\triangle WOS \sim \triangle GIP$, the perimeter of $\triangle GIP$ is $5x + 19$, and the perimeter of $\triangle WOS$ is 36.

5. Find the scale factor of $\triangle WOS$ to $\triangle GIP$

6. Find x.

7. Find the perimeter of $\triangle GIP$.

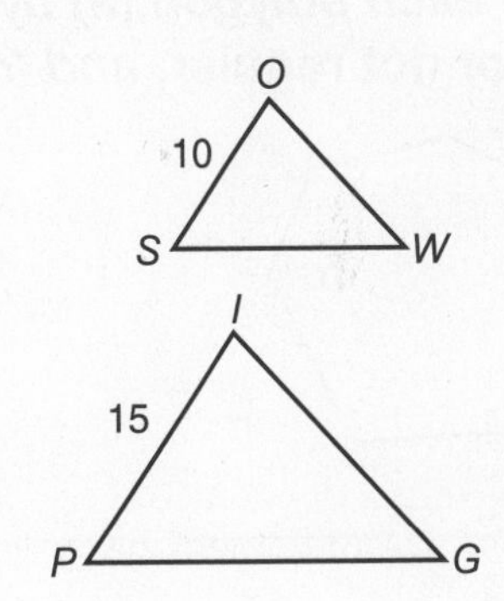

In the figure, $\triangle CAT \sim \triangle DOG$, and the perimeter of $\triangle CAT$ is 18.

8. Which side of $\triangle DOG$ corresponds to $\overline{AT}$ in $\triangle CAT$?

9. Write an expression for the perimeter of $\triangle DOG$.

10. Find x.

11. Find the perimeter of $\triangle DOG$.

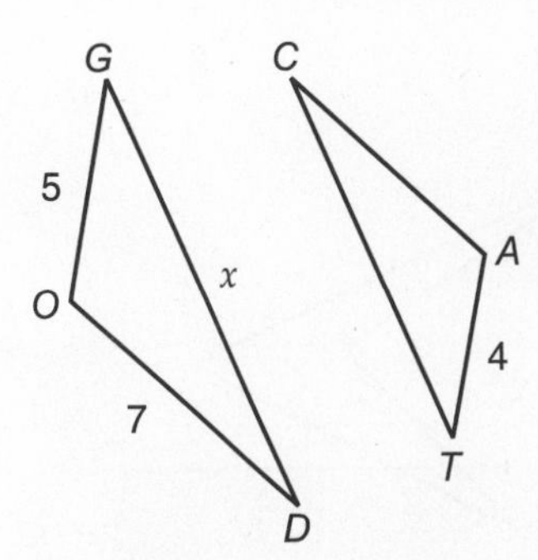

NAME ______________________ DATE ________ PERIOD ______

Study Guide

Student Edition
Pages 402–407

Naming Polygons

A **polygon** is a plane figure formed by a finite number of segments. In a **convex polygon**, all of the diagonals lie in the interior. A **regular polygon** is a convex polygon that is both equilateral and equiangular. In a **concave polygon**, any point of a diagonal lies in the exterior.

Example: Match each term with the appropriate letter from the box below. Some letters may be used more than once.

Term	Answer	Choice
1. concave polygon	c	**a.** *BVXKHW*
2. convex polygon	a	**b.** *AFNTR*
3. diagonal	e	**c.** *RFANT*
4. hexagon	a	**d.** $\overline{TR}$
5. pentagon	c	**e.** $\overline{BX}$
6. side	d	**f.** *R*

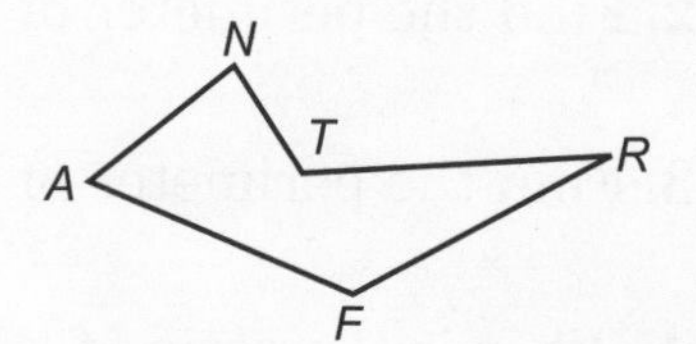

Classify each polygon (a) by the number of sides, (b) as regular ***or*** not regular, ***and (c) as*** convex ***or*** concave.

1.

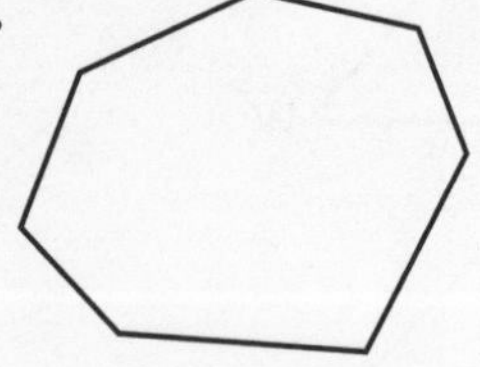

2.

3.

4.

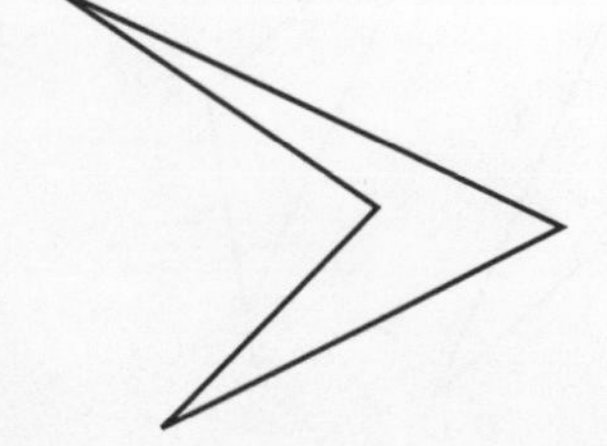

5.

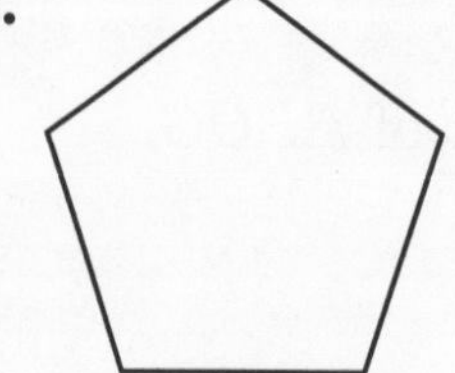

6.

10-2

NAME ______________________ DATE ________ PERIOD ______

Study Guide

Student Edition
Pages 408–412

Diagonals and Angle Measure

A **polygon** is a plane figure formed by a finite number of segments such that (1) sides that have a common endpoint are noncollinear and (2) each side intersects exactly two other sides, but only at their endpoints. A **convex polygon** is a polygon such that no line containing a side of the polygon contains a point in the interior of the polygon.

The following two theorems involve the interior and exterior angles of a convex polygon.

Interior Angle Sum Theorem	If a convex polygon has n sides, then the sum of the measures of its interior angles is $(n - 2)180$.
Exterior Angle Sum Theorem	In any convex polygon, the sum of the measures of the exterior angles, one at each vertex, is 360.

Example: Find the sum of the measures of the interior angles of a convex polygon with 13 sides.

$S = (n - 2)180$ *Interior Angle Sum Theorem*
$S = (13 - 2)180$
$S = (11)180$
$S = 1980$

Find the sum of the measures of the interior angles of each convex polygon.

1. 10-gon **2.** 16-gon **3.** 30-gon

The measure of an exterior angle of a regular polygon is given. Find the number of sides of the polygon.

4. 30 **5.** 20 **6.** 5

The number of sides of a regular polygon is given. Find the measures of an interior angle and an exterior angle for each polygon.

7. 18 **8.** 36 **9.** 25

10. The measure of the interior angle of a regular polygon is 157.5. Find the number of sides of the polygon.

NAME ______________________ DATE __________ PERIOD ______

Study Guide

Student Edition
Pages 413–418

Areas of Polygons

The following theorems involve the areas of polygons.

- For any polygon, there is a unique area.
- Congruent polygons have equal areas.
- The area of a given polygon equals the sum of the areas of the nonoverlapping polygons that form the given polygon.

Example: Find the area of the polygon in square units.

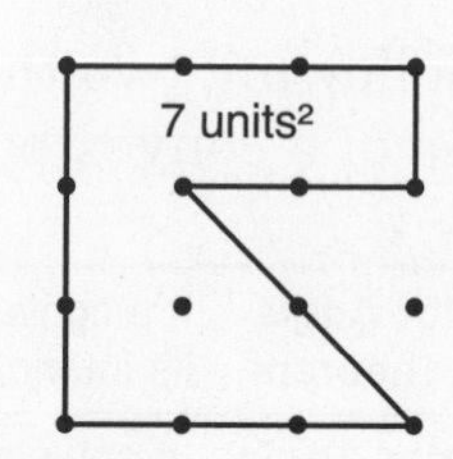

Since the area of each four-dot unit represents 1 square unit, the area of each three-dot unit represents 0.5 square unit.

$A = 6(1) + 2(0.5)$ *There are 6 four-dot units and 2 three-dot units.*

$A = 6 + 1$

$A = 7$

The area of the region is 7 square units, or 7 units^2.

Find the area of each polygon in square units.

1. 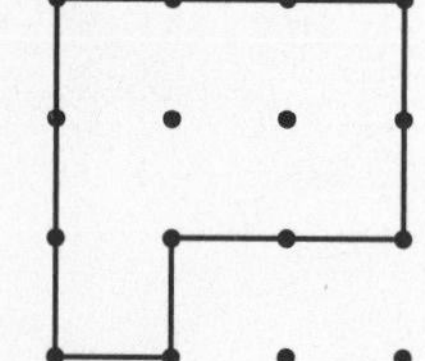

2. 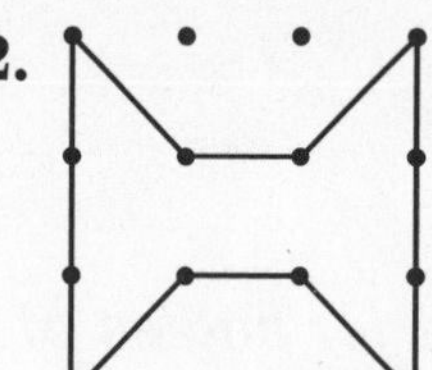

3. 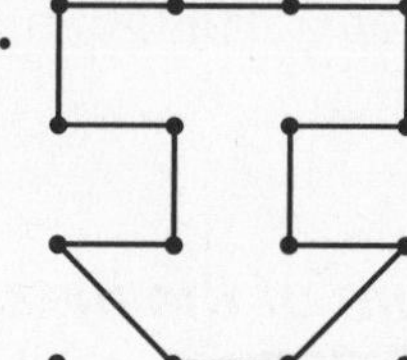

4. 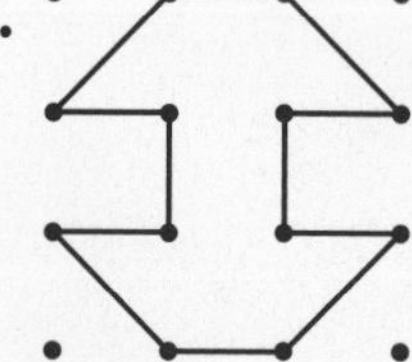

5. 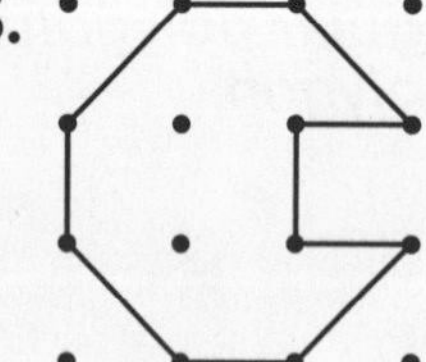

6. 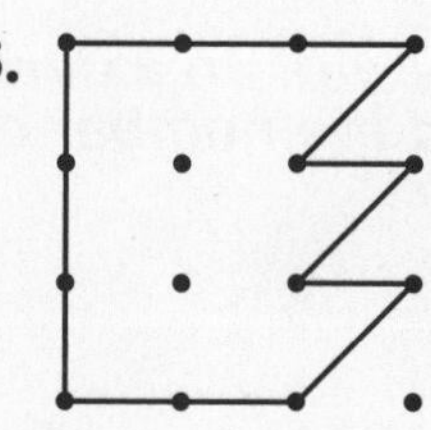

7. 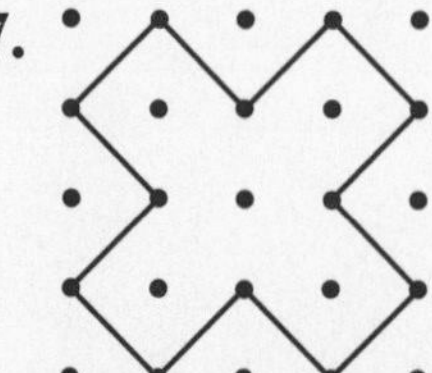

8. 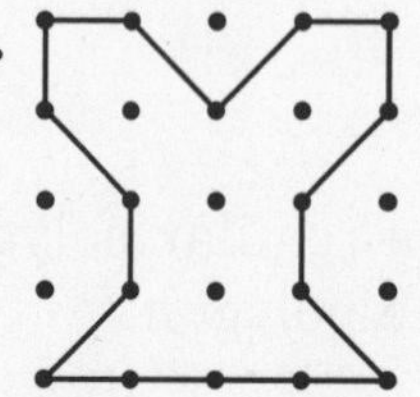

9. 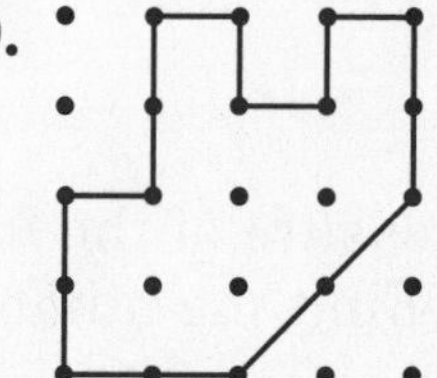

NAME ________________ DATE ________ PERIOD ______

Study Guide

Student Edition
Pages 419–424

Areas of Triangles and Trapezoids

The area of a triangle is equal to one-half its base times its altitude.
$A = \frac{1}{2}bh$

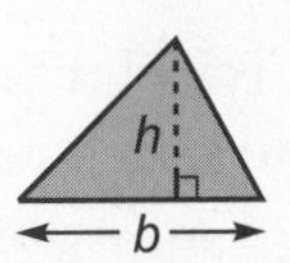

Example 1: Find the area of the triangle.

$A = \frac{1}{2}bh$

$A = \frac{1}{2}(7.5)(4)$ $\quad b = 7.5, h = 4$

$A = 2(7.5)$

$A = 15$ The area is 15 square centimeters.

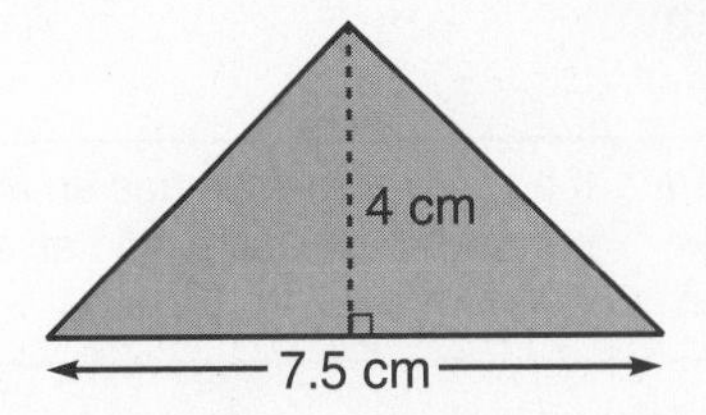

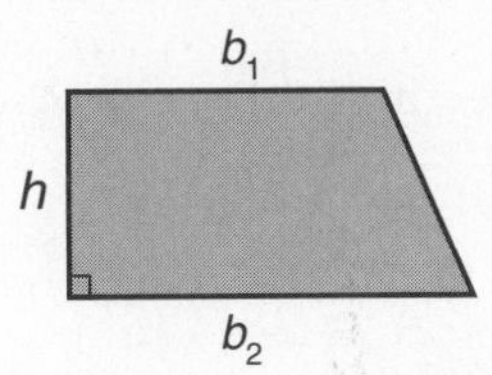

The area of a trapezoid is equal to one-half its altitude times the sum of its bases: $A = \frac{1}{2}h(b_1 + b_2)$.

Example 2: Find the area of the trapezoid.

$A = \frac{1}{2}h(b_1 + b_2)$

$A = \frac{1}{2}(6)\left(10\frac{1}{2} + 14\frac{1}{2}\right)$ $\quad h = 6, b_1 = 10\frac{1}{2}, b_2 = 14\frac{1}{2}$

$A = 3(25)$

$A = 75$ The area is 75 square inches.

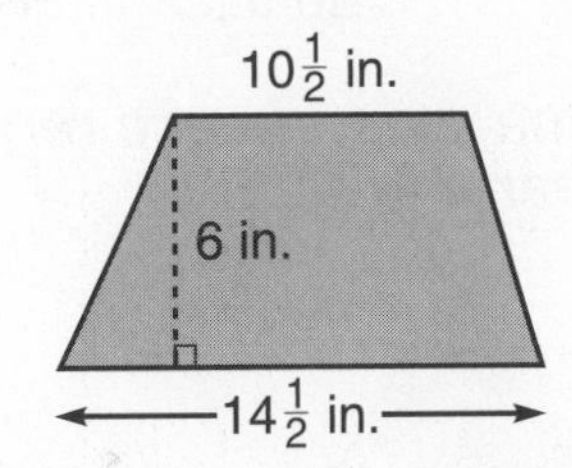

Find the area of each triangle or trapezoid.

1.

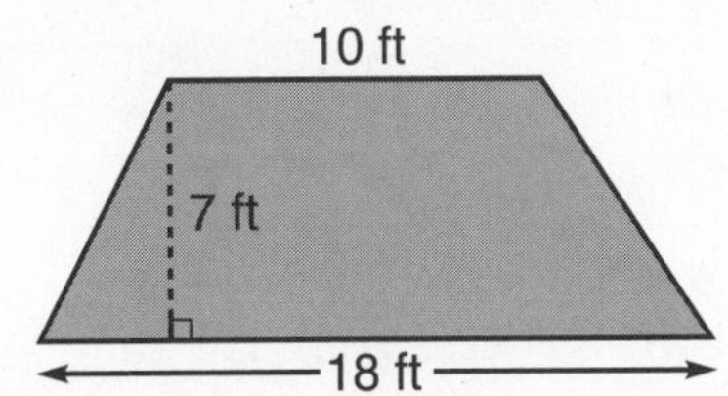

2.

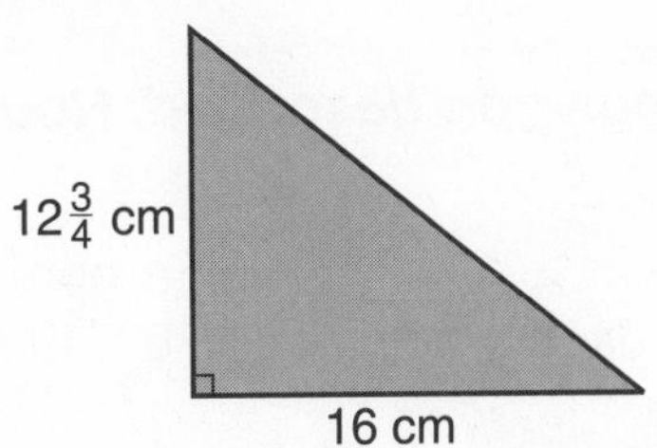

3.

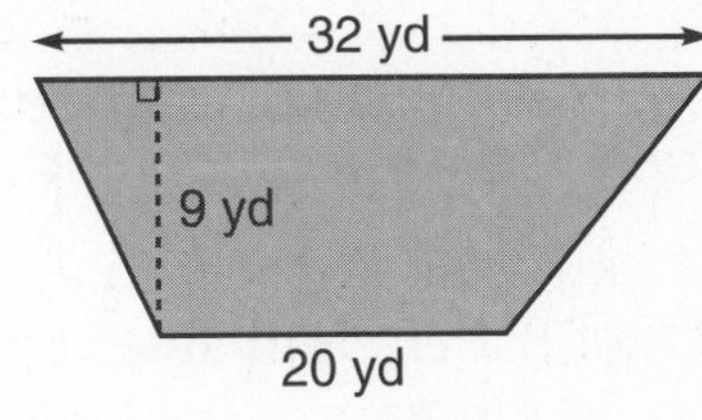

4.

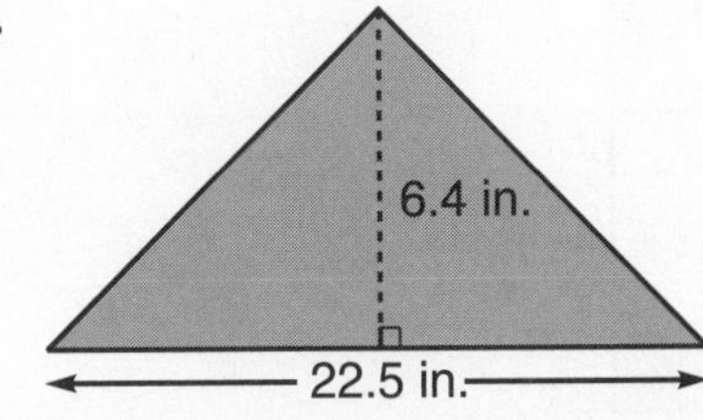

5.

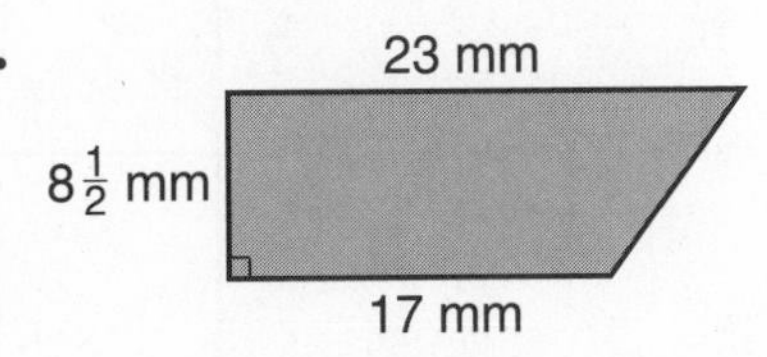

6.

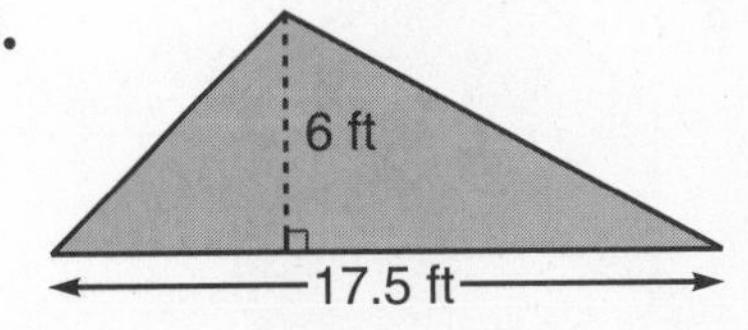

NAME ______________________________ DATE __________ PERIOD _______

Study Guide

Student Edition
Pages 425–431

Areas of Regular Polygons

In a regular polygon, a segment drawn from the center of the polygon perpendicular to a side of the polygon is called an **apothem**. In the figure at the right, $\overline{PS}$ is an apothem.

Area of a Regular Polygon	If a regular polygon has an area of A square units, a perimeter of P units, and an apothem of a units, then $A = \frac{1}{2}Pa$.

Example: Find the area of a regular pentagon with an apothem of 2.8 cm and a perimeter of 20.34 cm.

$A = \frac{1}{2}Pa$ *Area of a regular polygon*

$A = \frac{1}{2}(20.34)(2.8)$

$A = 28.476$

The area is 28.476 square centimeters.

Find the area of each regular polygon. Round your answers to the nearest tenth.

1.

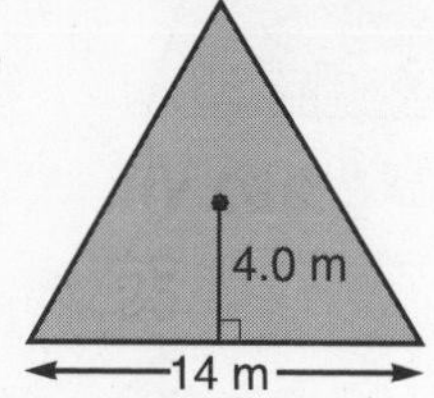

2.

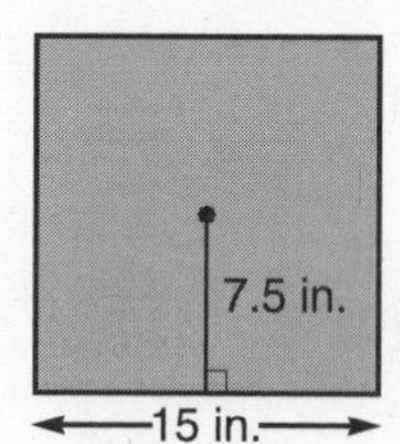

3.

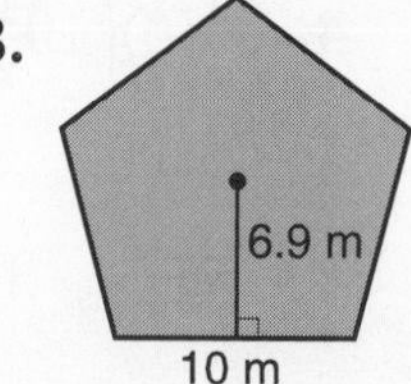

Find the area of each regular polygon described. Round your answers to the nearest tenth.

4. a hexagon with an apothem of 8.7 cm and sides that are each 10 cm long

5. a pentagon with a perimeter of 54.49 m and an apothem of 7.5 m

Find the area of each shaded region.

6.

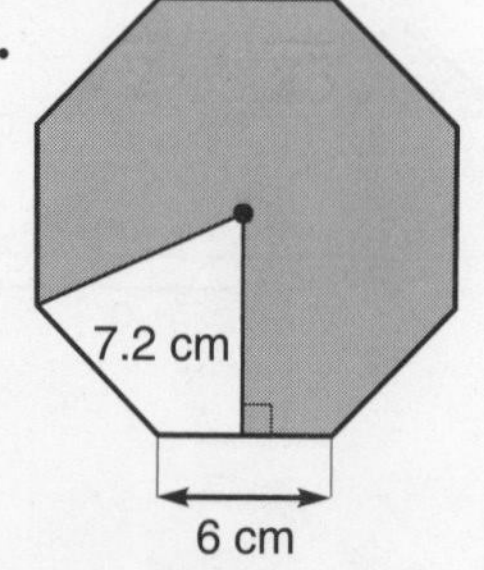

7.

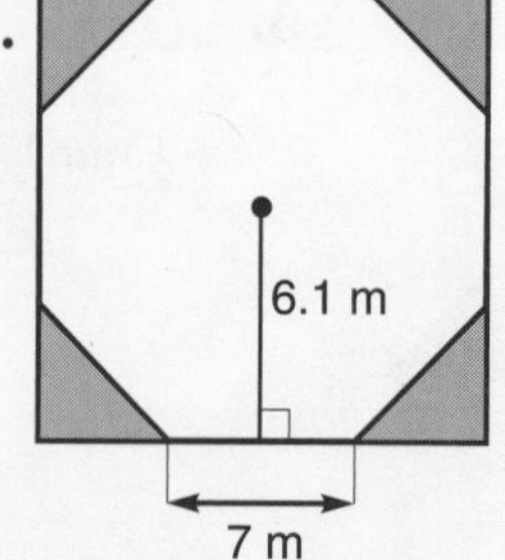

Study Guide

Symmetry

If you can fold a figure along a line so that the two parts reflect each other, the figure has a line of symmetry.

Examples one line symmetry three lines of symmetry no lines of symmetry

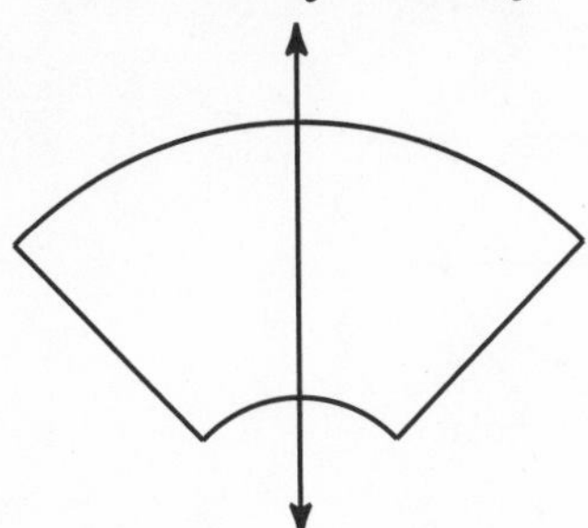

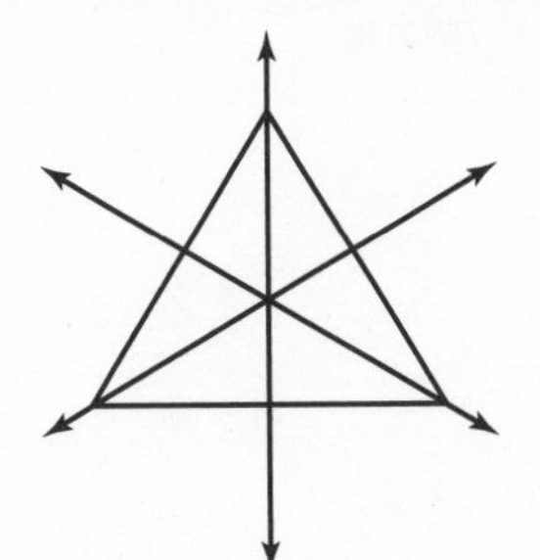

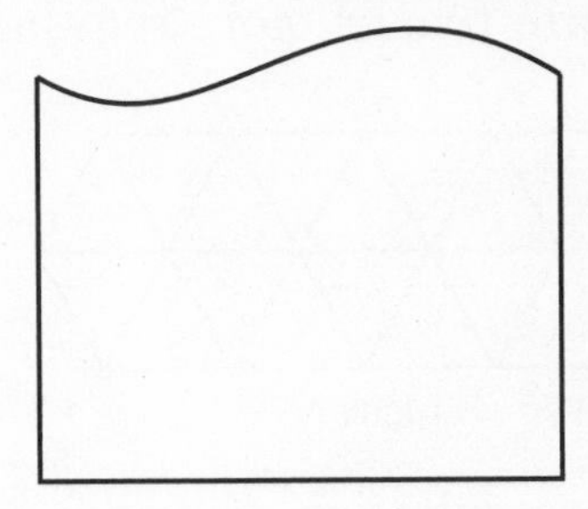

A figure has rotational symmetry if it can be turned less than 360° about its center and it looks like the original figure.

Example Original 72° turn 144° turn 216° turn 288° turn

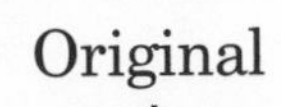

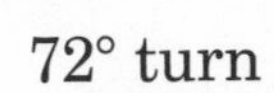

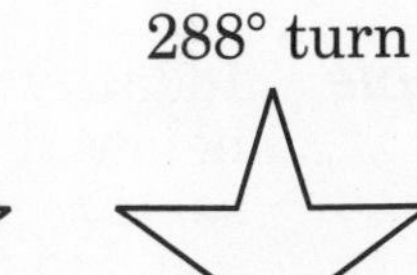

Draw the line(s) of symmetry for each figure.

1.

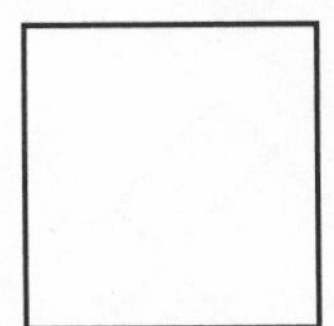

2.

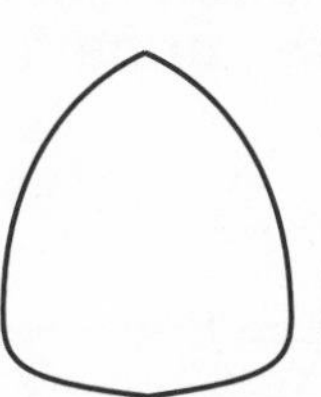

3.

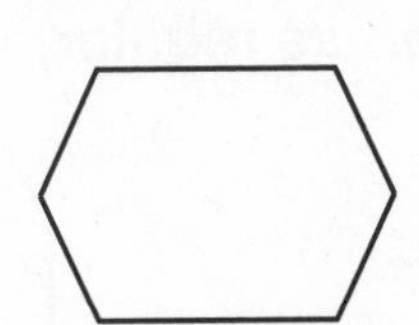

4.

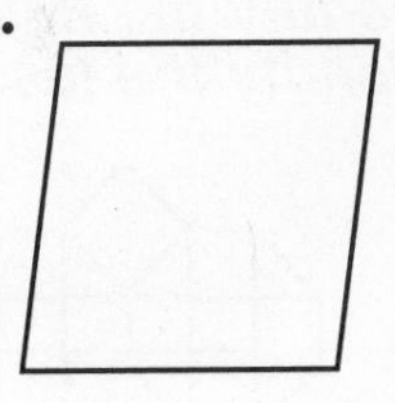

Determine whether each figure has rotational symmetry.

5.

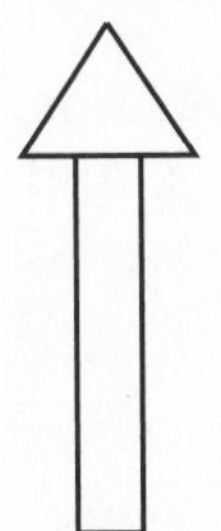

6.

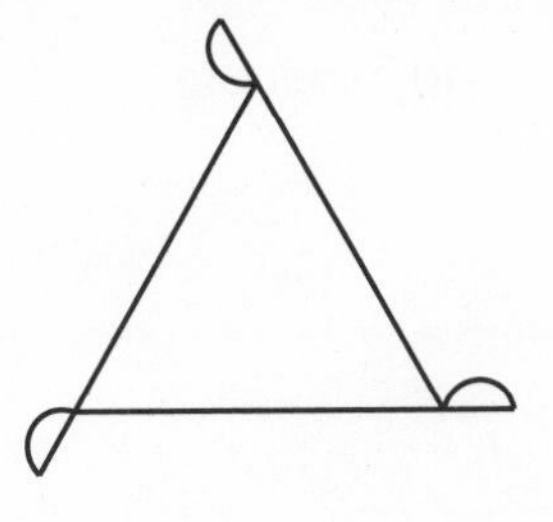

7.

8.

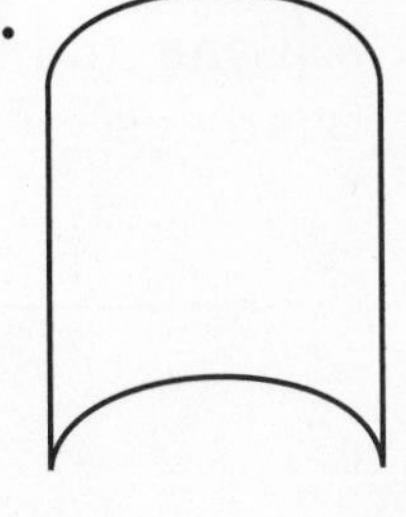

NAME ______________________ DATE __________ PERIOD ______

Study Guide

Tessellations

Tessellations are patterns that cover a plane with repeating polygons so that there are no overlapping or empty spaces. A **regular tessellation** uses only one type of regular polygon. **Semi-regular tessellations** are uniform tessellations that contain two or more regular polygons.

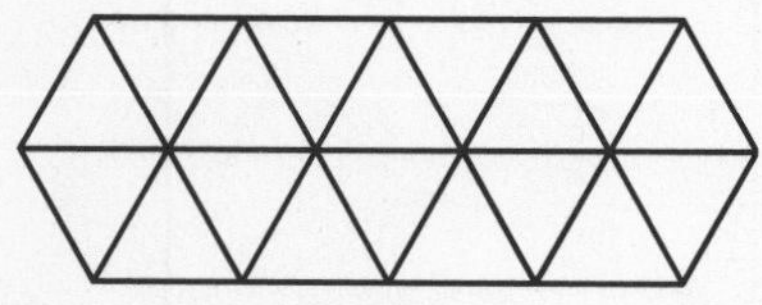
regular

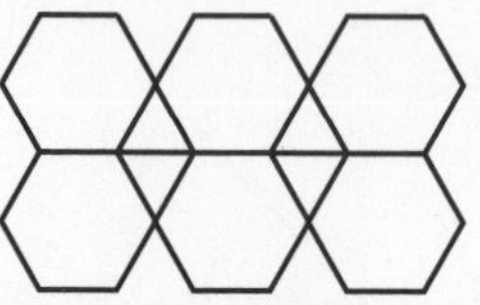
semi-regular

In a tessellation, the sum of the measures of the angles of the polygons surrounding a point (at a vertex) is 360. If a regular polygon has an interior angle with a measure that is a factor of 360, then the polygon will tessellate.

Example: Identify the figures used to create the tessellation below. Then identify the tessellation as *regular*, *semi-regular*, or *neither.*

Regular pentagons and two different types of parallelograms are used. Since the parallelograms are not regular, this tessellation is neither regular nor semi-regular.

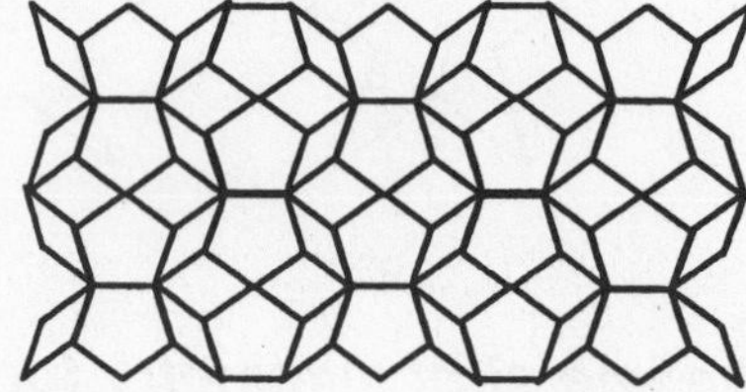

***Identify the figures used to create each tessellation. Then identify the tessellation as* regular, semi-regular, *or* neither.**

1.

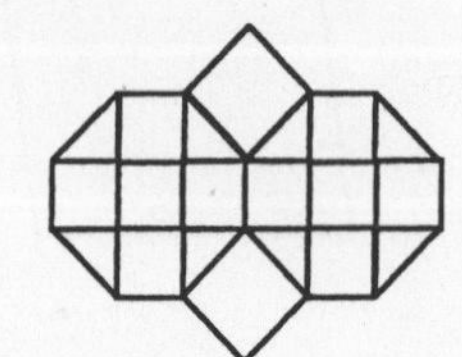

2.

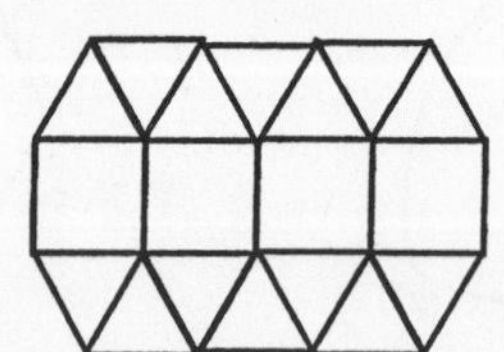

3.

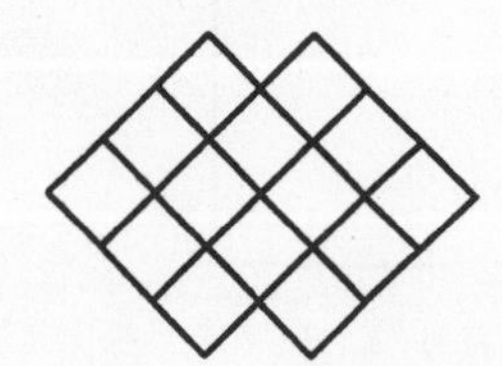

Use isometric or rectangular dot paper to create a tessellation using the given polygons.

4. scalene and right triangles

5. squares and isosceles trapezoids

6. parallelograms, rectangles, and hexagons

NAME ______________________ DATE ________ PERIOD ______

Study Guide

Student Edition
Pages 454–459

Parts of a Circle

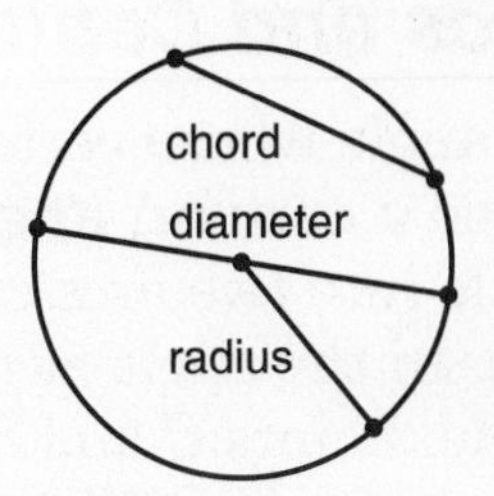

A **circle** is the set of all points in a plane that are a given distance from a given point in the plane called the **center**. Various parts of a circle are labeled in the figure at the right. Note that the diameter is twice the radius.

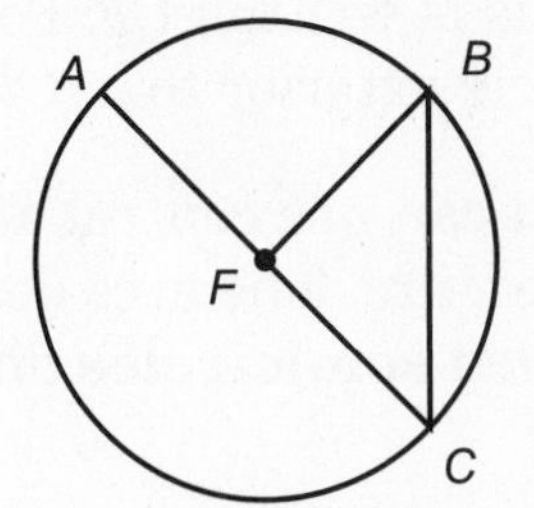

Example: In $\odot F$, $\overline{AC}$ is a diameter.

- Name the circle. $\odot F$
- Name a radius. $\overline{AF}$, $\overline{CF}$, or $\overline{BF}$
- Name a chord that is not a diameter. $\overline{BC}$

Use ⊙S to name each of the following.

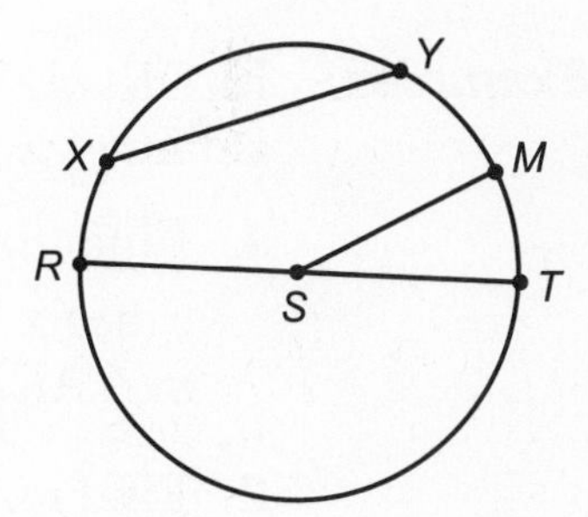

1. the center

2. three radii

3. a diameter

4. a chord

Use ⊙P to determine whether each statement is true or false.

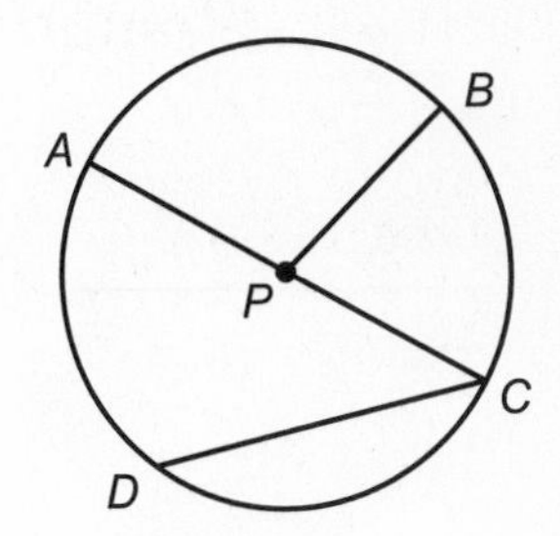

5. $\overline{PC}$ is a radius of $\odot P$.

6. $\overline{AC}$ is a chord of $\odot P$.

7. If $PB = 7$, then $AC = 14$.

On a separate sheet of paper, use a compass and a ruler to make a drawing that fits each description.

8. $\odot A$ has a radius of 2 inches. $\overline{QR}$ is a diameter.

9. $\odot G$ has a diameter of 2 inches. Chord $\overline{BC}$ is 1 inch long.

NAME ______________________ DATE ________ PERIOD ______

Study Guide

Student Edition
Pages 462–467

Arcs and Central Angles

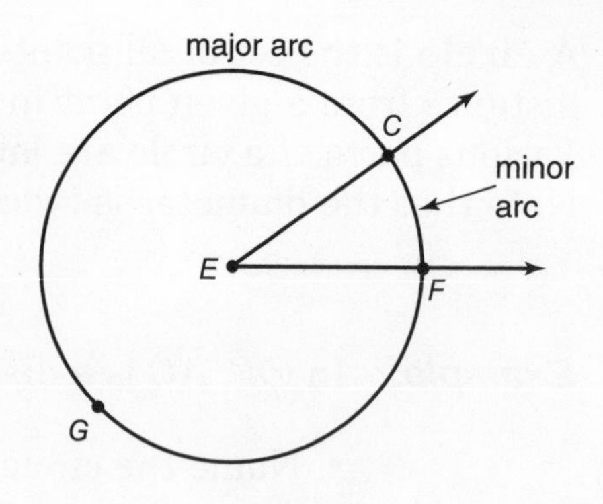

An angle whose vertex is at the center of a circle is called a **central angle**. A central angle separates a circle into two arcs called a **major arc** and a **minor arc**. In the circle at the right, $\angle CEF$ is a central angle. Points C and F and all points of the circle interior to $\angle CEF$ form a minor arc called arc CF. This is written $\overparen{CF}$. Points C and F and all points of the circle exterior to $\angle CEF$ form a major arc called $\overparen{CGF}$.

You can use central angles to find the degree measure of an arc. The arcs determined by a diameter are called **semicircles** and have measures of 180.

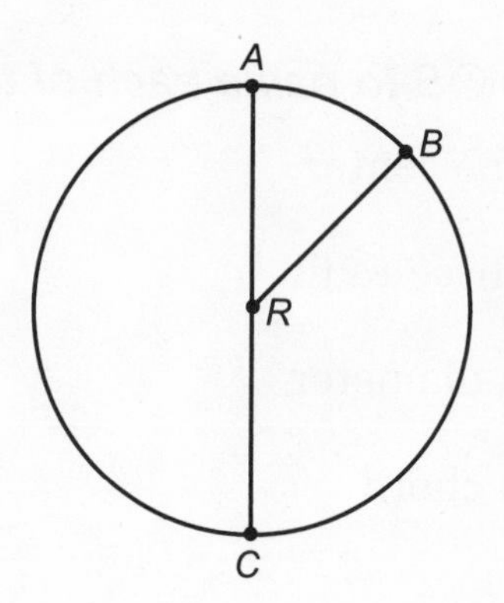

Examples: In $\odot R$, $m\angle ARB = 42$ and $\overline{AC}$ is a diameter.

1 Find $m\overparen{AB}$.

Since $\angle ARB$ is a central angle and $m\angle ARB = 42$, then $m\overparen{AB} = 42$.

2 Find $m\overparen{ACB}$.

$m\overparen{ACB} = 360 - m\angle ARB = 360 - 42$ or 318

3 Find $m\overparen{CAB}$.

$$\begin{aligned} m\overparen{CAB} &= m\overparen{ABC} + m\overparen{AB} \\ &= 180 + 42 \\ &= 222 \end{aligned}$$

Refer to $\odot P$ for Exercises 1–4. If $\overline{SN}$ and $\overline{MT}$ are diameters with $m\angle SPT = 51$ and $m\angle NPR = 29$, determine whether each arc is a minor arc, a major arc, or a semicircle. Then find the degree measure of each arc.

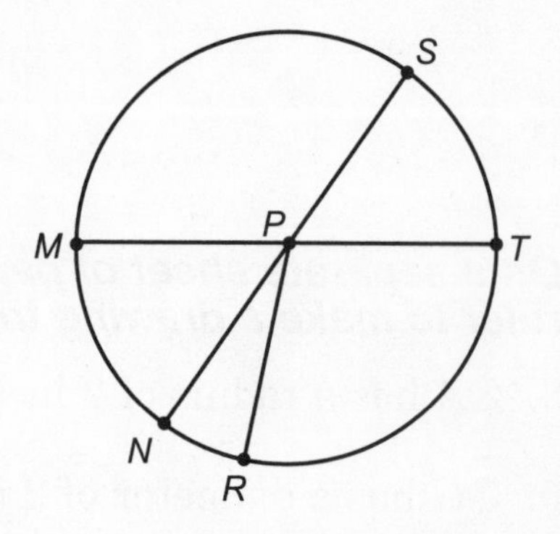

1. $m\overparen{NR}$ **2.** $m\overparen{ST}$

3. $m\overparen{TSR}$ **4.** $m\overparen{MST}$

11-3

NAME ______________________ DATE __________ PERIOD ______

Study Guide

Student Edition
Pages 468–473

Arcs and Chords

The following theorems state relationships between arcs, chords, and diameters.

- In a circle or in congruent circles, two minor arcs are congruent if and only if their corresponding chords are congruent.
- In a circle, a diameter bisects a chord and its arc if and only if it is perpendicular to the chord.

Example: In the circle, O is the center, $OD = 15$, and $CD = 24$. Find x.

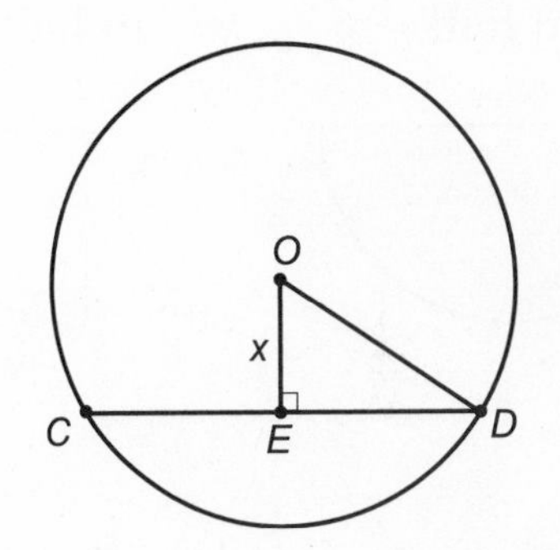

$$ED = \frac{1}{2}CD$$
$$= \frac{1}{2}(24)$$
$$= 12$$

$$(OE)^2 + (ED)^2 = (OD)^2$$
$$x^2 + 12^2 = 15^2$$
$$x^2 + 144 = 225$$
$$x^2 = 81$$
$$x = 9$$

In each circle, O is the center. Find each measure.

1. $m\widehat{NP}$

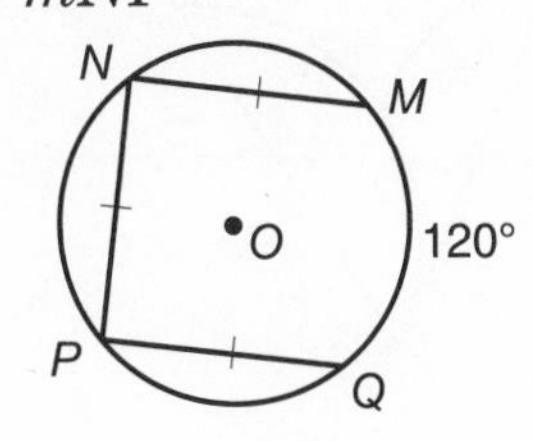

2. KM

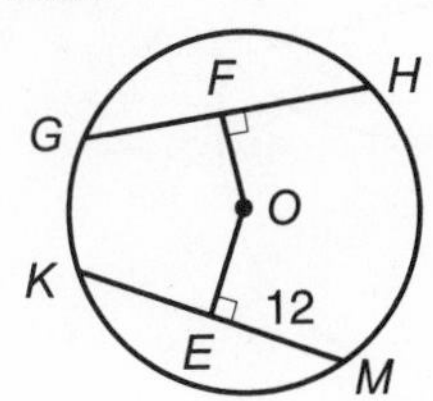

3. XY

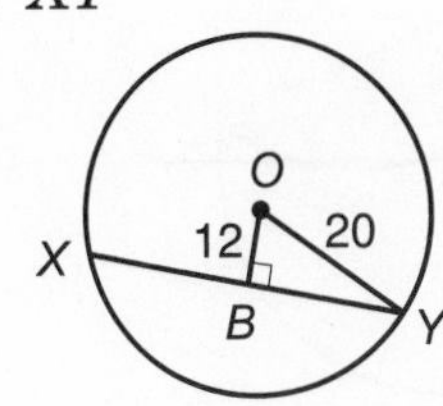

4. Suppose a chord is 20 inches long and is 24 inches from the center of the circle. Find the length of the radius.

5. Suppose a chord of a circle is 5 inches from the center and is 24 inches long. Find the length of the radius.

6. Suppose the diameter of a circle is 30 centimeters long and a chord is 24 centimeters long. Find the distance between the chord and the center of the circle.

NAME ______________________ DATE __________ PERIOD _____

Study Guide

Student Edition
Pages 474–477

Inscribed Polygons

You can make many regular polygons by folding a circular piece of paper. The vertices of the polygon will lie on the circle, so the polygon is said to be inscribed in the circle.

1. Draw a circle with a radius of 2 inches and cut it out. Make the following folds to form a square.

Step A
Fold the circle in half.

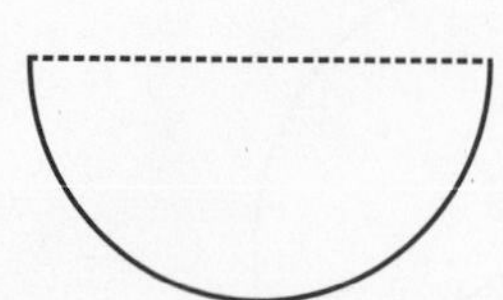

Step B
Fold the circle in half again.

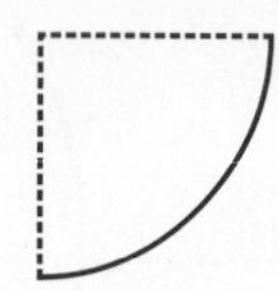

Step C
Unfold the circle.

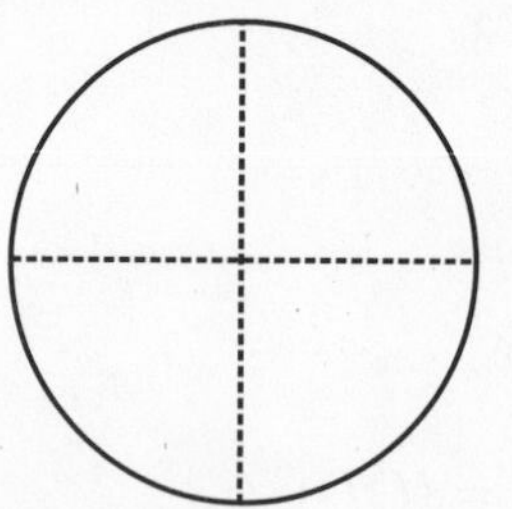

Step D
Fold the four arcs designated by the creases.

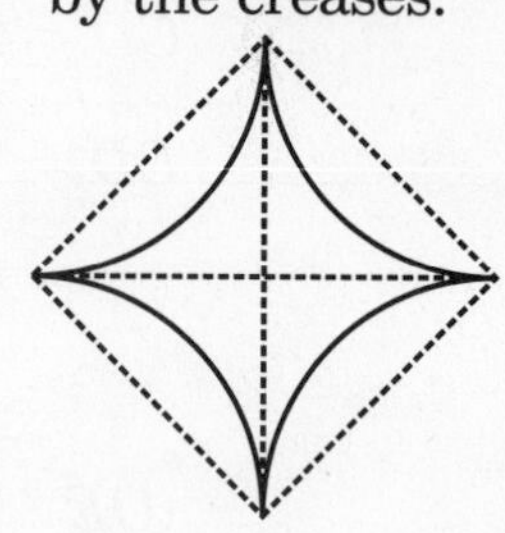

2. Draw another circle with a radius of 2 inches and cut it out. Make the following folds to form a regular triangle.

Step A
Fold one portion in toward the center.

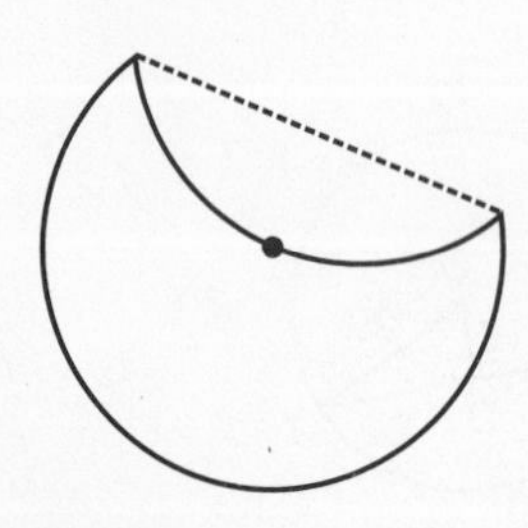

Step B
Fold another portion in toward the center, overlapping the first.

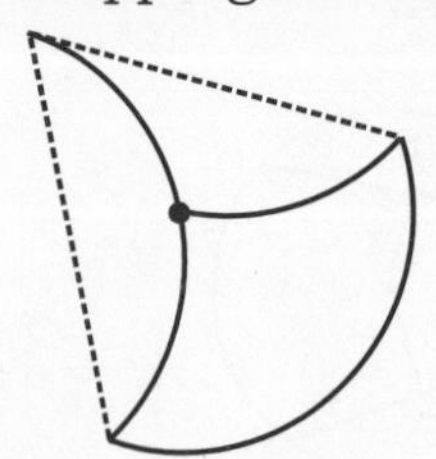

Step C
Fold the remaining third of the circle in toward the center.

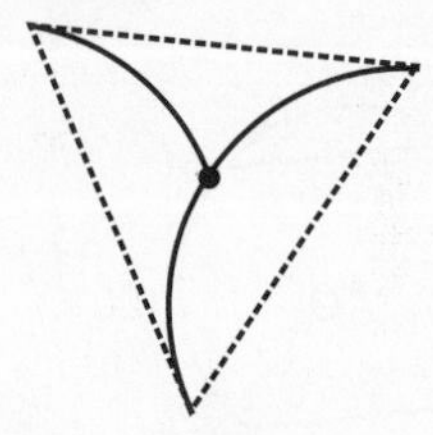

3. Cut out another circle and fold it to make a regular octagon. Draw the steps used.

4. Cut out another circle and fold it to make a regular hexagon. Draw the steps used.

5. Cut out a circle with radius 4 inches and fold it to make a regular dodecagon. Draw the steps used.

NAME ______________________ DATE ________ PERIOD ______

11-5 Study Guide

Student Edition
Pages 478–482

Circumference of a Circle

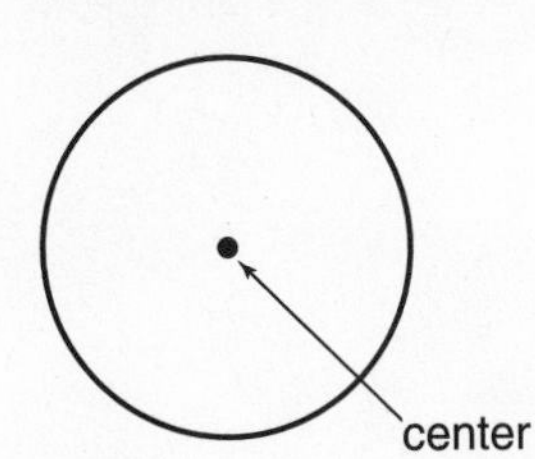

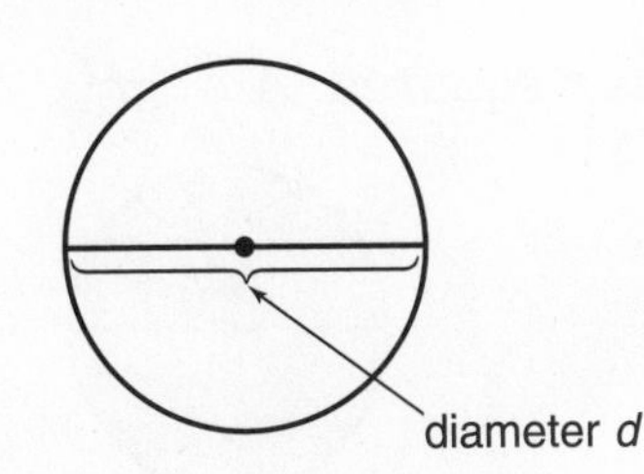

radius r

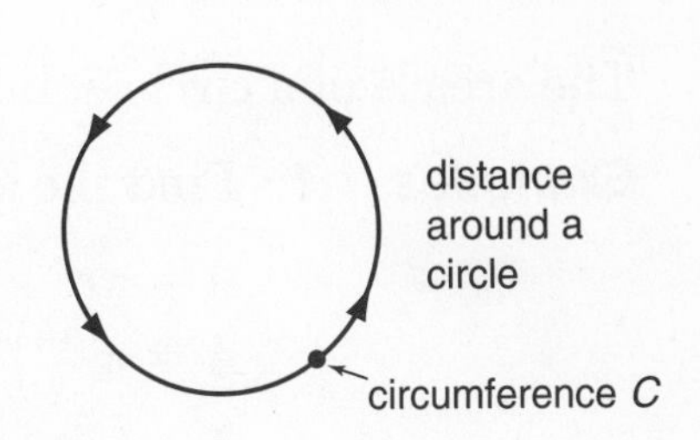

Examples: Find the circumference of each circle. Use 3.14 for π.

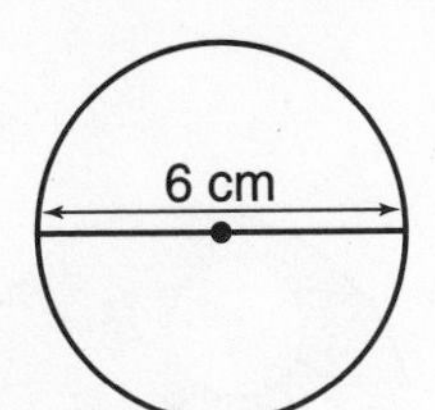

$C = \pi d$
$C = \pi(6)$
$C \approx 18.85$
$C \approx 19$ cm

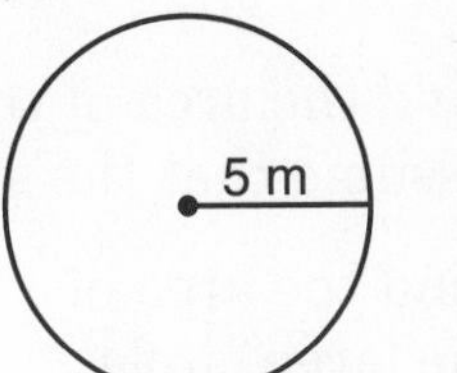

$C = 2\pi r$
$C = 2\pi(5)$
$C = 10\pi$
$C \approx 31.4$
$C \approx 31$ m

Find the circumference of each circle.

1.

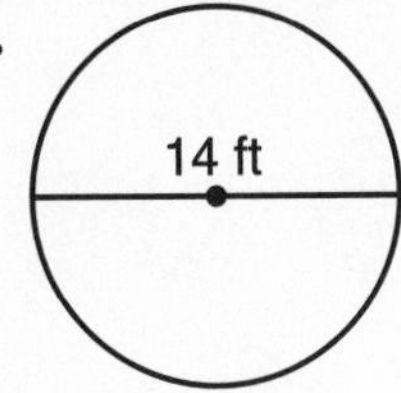

2.

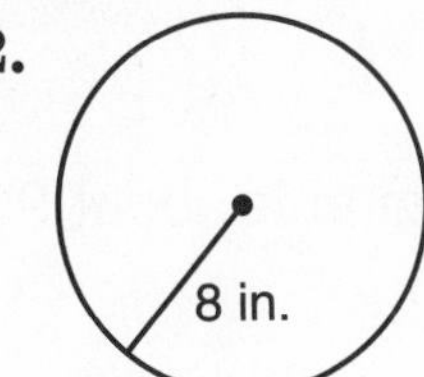

3.

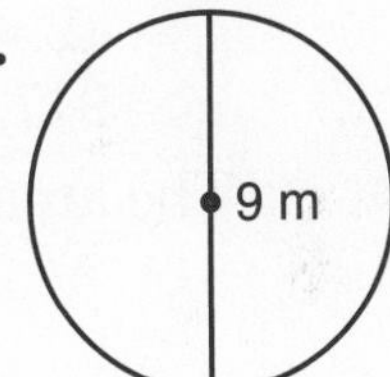

4. The radius is $6\frac{1}{5}$ feet.

5. The diameter is 4.7 yards.

Solve. Round to the nearest inch.

6. What is the circumference of the top of an ice cream cone if its diameter is about $1\frac{7}{8}$ inches?

7. The radius of the basketball rim is 9 inches. What is the circumference?

NAME ______________________ DATE __________ PERIOD _____

Study Guide

Student Edition
Pages 483–487

Area of a Circle

The area A of a circle equals π times the radius r squared: $A = \pi r^2$.

Examples **1** Find the area of the circle.

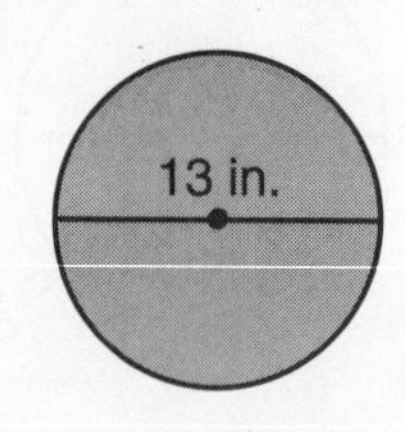

$A = \pi r^2$
$A = \pi\left(\frac{13}{2}\right)^2$
$A = \pi(42.25)$
$A \approx 132.73$
The area of the circle is about 132.7 in^2.

2 Find the area of the shaded region.
Assume that the smaller circles are congruent.

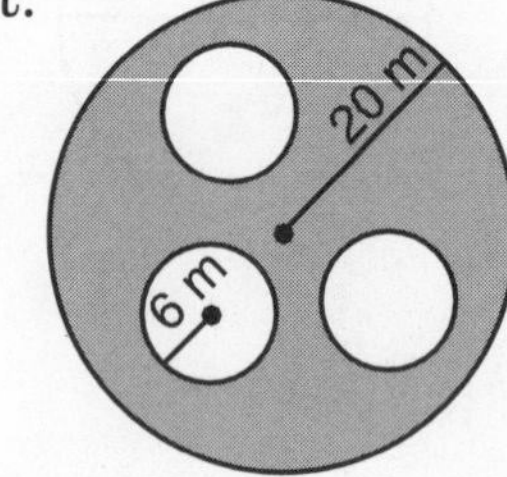

Find the area of the large circle.

$A = \pi r^2$
$A = \pi(20)^2$
$A \approx 1256.64$

Find the area of a small circle.

$A = \pi r^2$
$A = \pi(6)^2$
$A \approx 113.10$

Now find the area of the shaded region.
$A \approx 1256.64 - 3(113.10)$
$\approx 1256.64 - 339.3$
≈ 917.34
The area of the shaded region is about 917.3 m^2.

Find the area of each circle to the nearest tenth.

1. 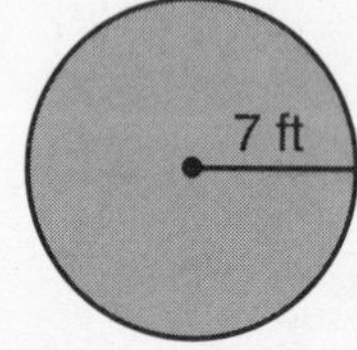

2. 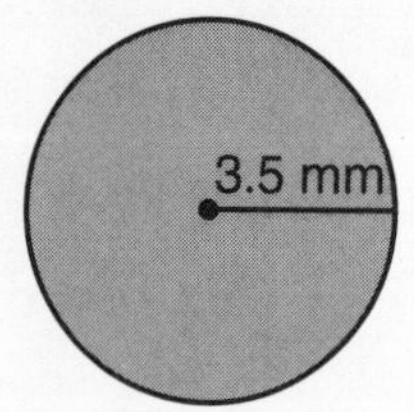

3. 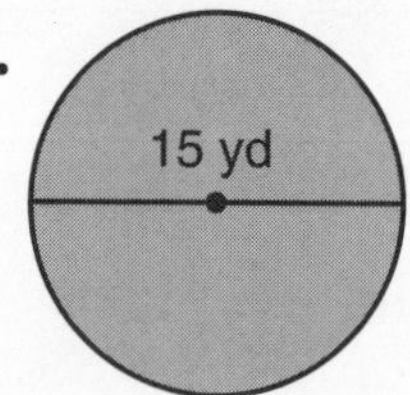

Find the area of each shaded region to the nearest tenth.

4. 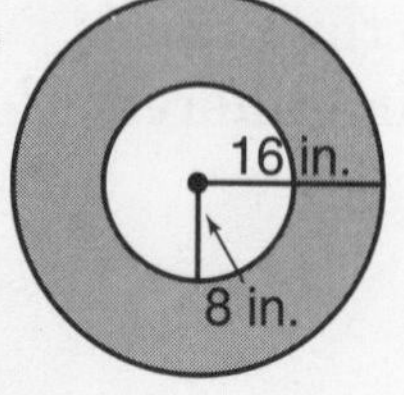

5. 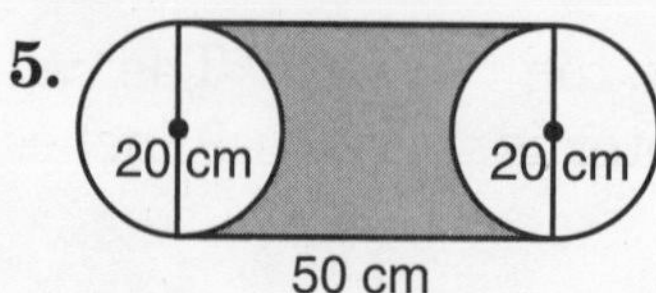

6. 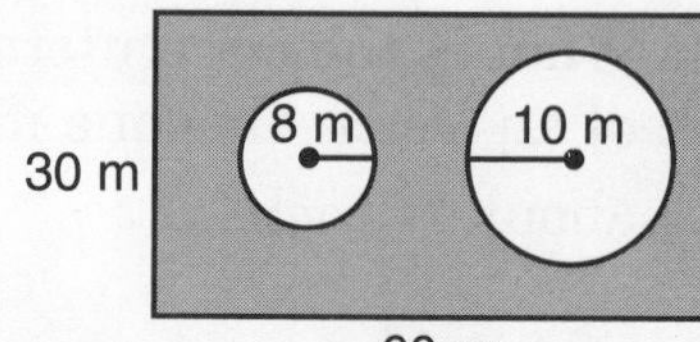

NAME ______________________ DATE __________ PERIOD ______

Study Guide

Student Edition
Pages 496–501

Solid Figures

Prisms have two parallel faces, called **bases**, that are congruent polygons. The other faces are called **lateral faces**. **Pyramids** have a polygon for a base and triangles for sides. Prisms and pyramids are named by the shape of their bases.

Example: Use isometric dot paper to sketch a hexagonal prism that is 5 units long.

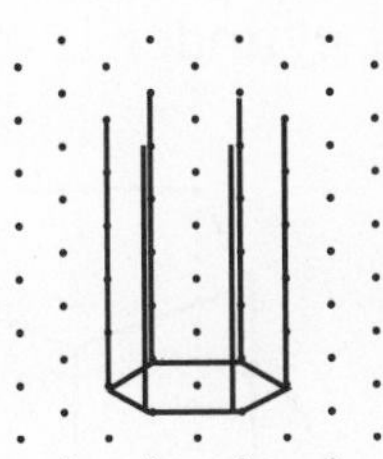

Step 1 Lightly draw a hexagon for a base.

Step 2 Lightly draw the vertical segments at the vertices of the base. Each segment is 5 units high.

Step 3 Complete the top of the prism.

Step 4 Go over your lines. Use dashed lines for the edges of the prism you cannot see from your perspective and solid lines for the edges you can see.

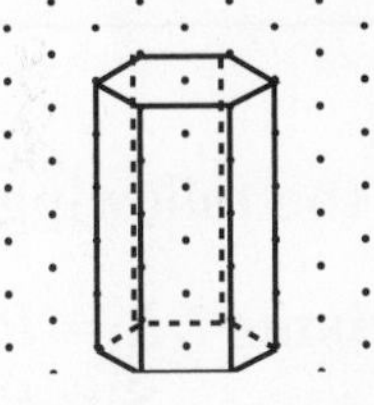

Use isometric dot paper to draw each solid.

1. a rectangular prism that is 2 units high, 5 units long, and 3 units wide

2. a pentagonal prism that is 3 units high

3. a square pyramid with a base that is 4 units wide

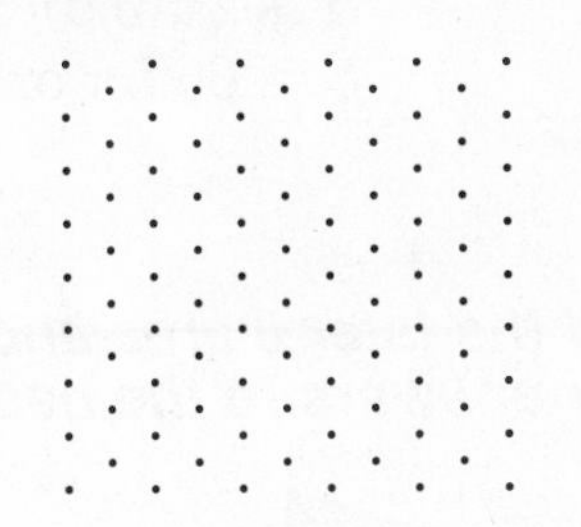

Name each solid.

4.

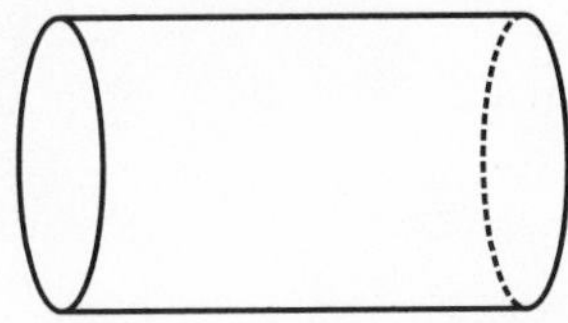

5.

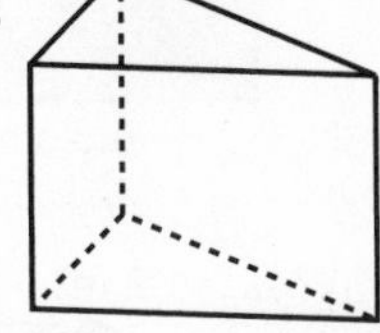

6.

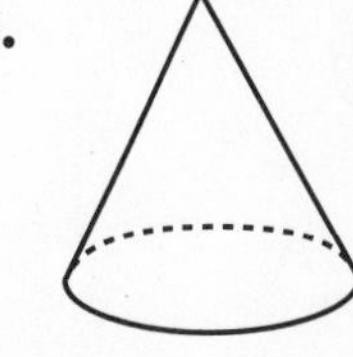

NAME ______________________________ DATE __________ PERIOD _____

Study Guide

Student Edition
Pages 504–509

Surface Areas of Prisms and Cylinders

Prisms are polyhedrons with congruent polygonal bases in parallel planes. **Cylinders** have congruent and parallel circular bases. An **altitude** is a perpendicular segment joining the planes of the bases. The length of an altitude is the **height** of the figure. **Right prisms** have lateral edges that are altitudes. A right cylinder is one whose **axis** is an altitude.

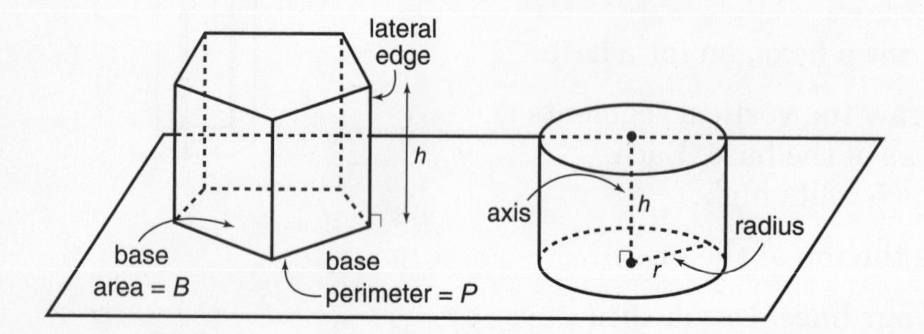

In the following formulas, L is lateral and S is surface area.

Prisms $L = Ph$
$S = Ph + 2B$

Cylinders $L = 2\pi rh$
$S = 2\pi rh + 2\pi r^2$

Example: Find the surface area of the cylinder.

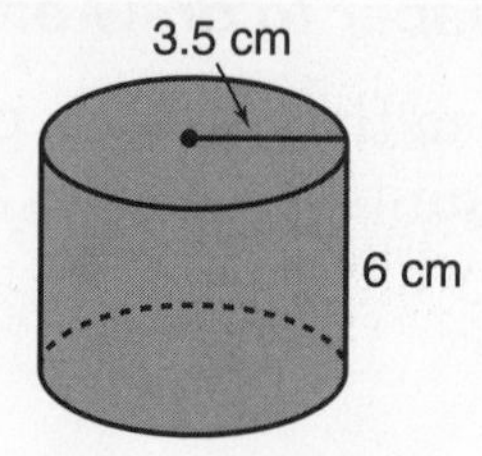

$S = 2\pi rh + 2\pi r^2$
$S = 2\pi(3.5)(6) + 2\pi(3.5)^2$
$S = 66.5\pi$ or about 208.92 cm^2

Find the lateral area and the surface area of each solid. Round your answers to the nearest tenth, if necessary.

1.

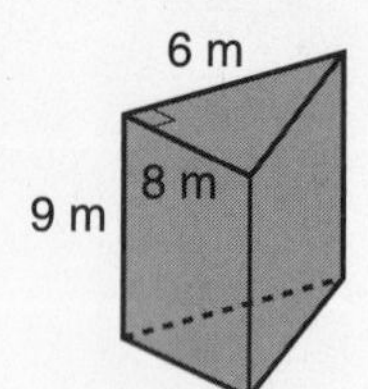

2.

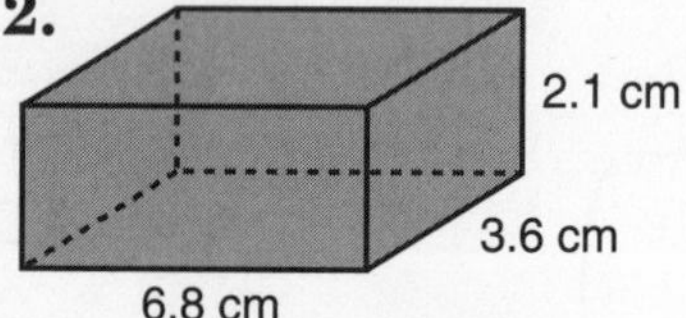

3.

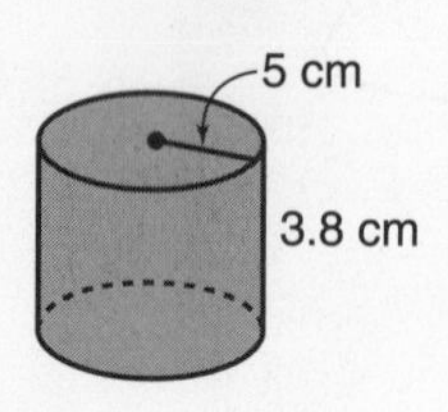

4.

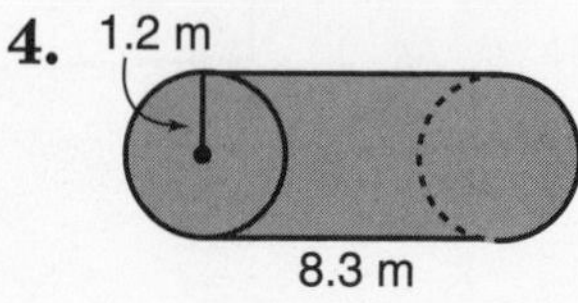

NAME ______________________ DATE __________ PERIOD _______

Study Guide

Student Edition
Pages 510–515

Volumes of Prisms and Cylinders

The measure of the amount of space that a figure encloses is the **volume** of the figure. Volume is measured in cubic units such as cubic yards or cubic feet. A cubic foot is equivalent to a cube that is 1 foot long on each side. A cubic yard is equivalent to 27 cubic feet.

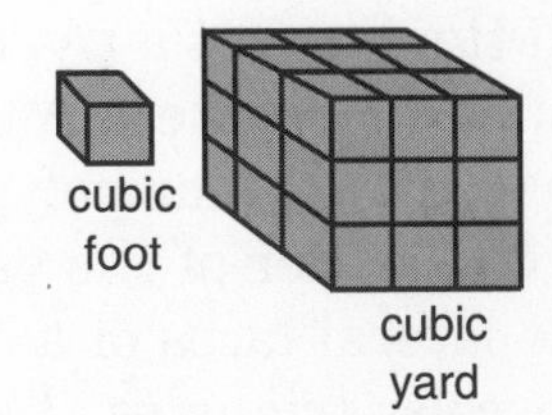

Volume of a Prism	If a prism has a volume of V cubic units, a base with an area of B square units, and a height of h units, then $V = Bh$.
Volume of a Cylinder	If a cylinder has a volume of V cubic units, a height of h units, and a radius of r units, then $V = \pi r^2 h$.

Examples: Find the volume of each solid.

1

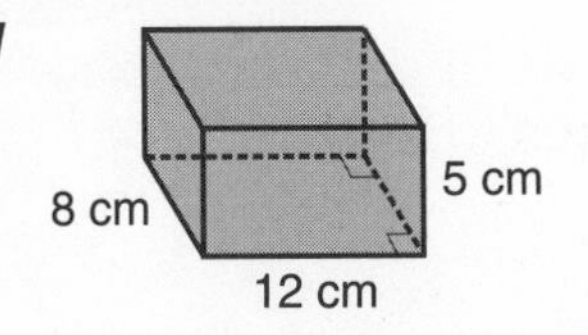

$V = Bh$
$V = (8)(12)(5)$
$V = 480 \text{ cm}^3$

2

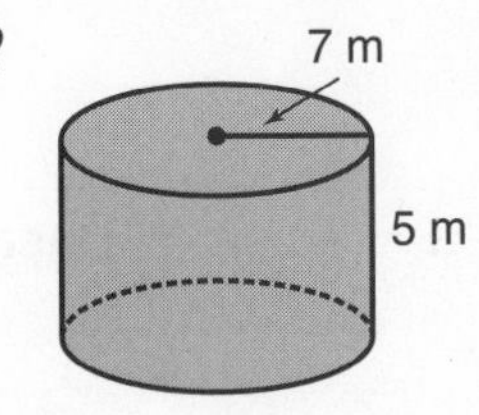

$V = \pi r^2 h$
$V = \pi(7)^2(5)$
$V = 245\pi$ or about 769.7 m^3

Find the volume of each solid. Round to the nearest hundredth, if necessary.

1.

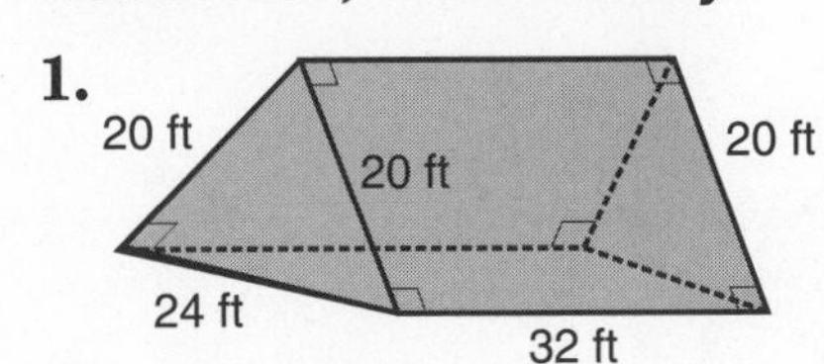

2.

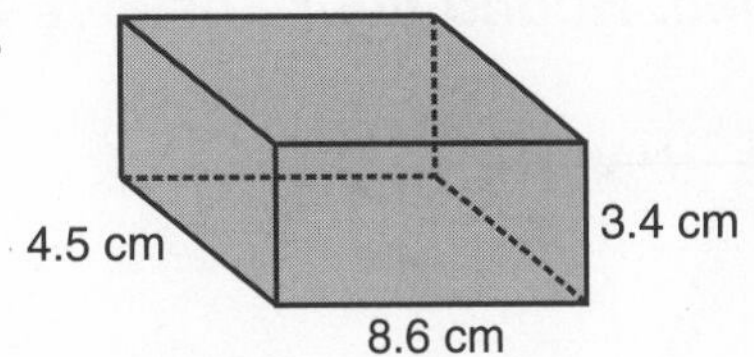

3.

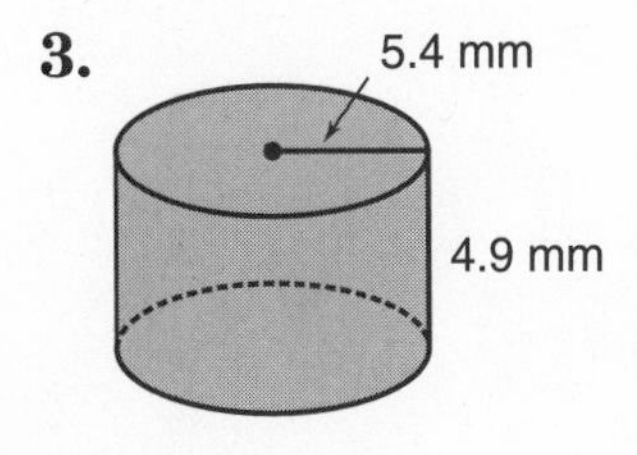

4.

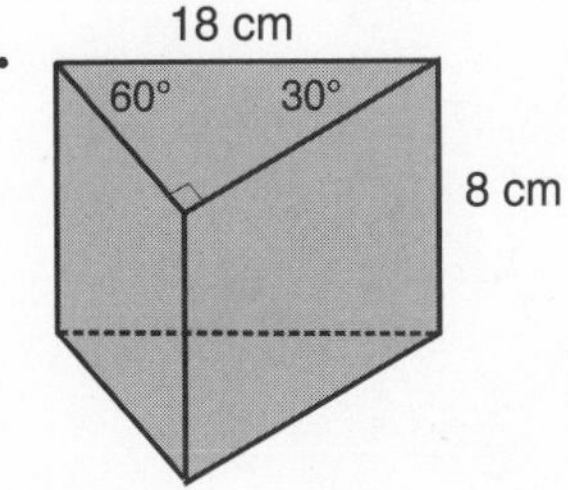

NAME ______________________________ DATE __________ PERIOD _____

Study Guide

Surface Areas of Pyramids and Cones

All the faces of a **pyramid**, except one, intersect at a point called the **vertex**. A pyramid is a **regular pyramid** if its base is a regular polygon and the segment from the vertex to the center of the base is perpendicular to the base. All the lateral faces of a regular pyramid are congruent isosceles triangles. The height of each lateral face is called the **slant height**.

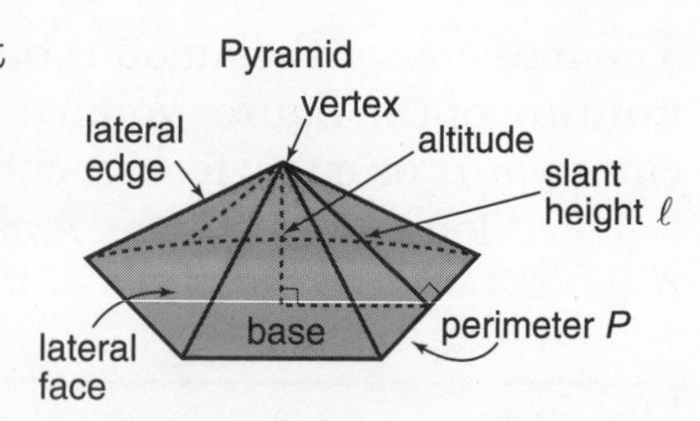

The slant height of a right circular cone is the length of a segment from the vertex to the edge of the circular base.

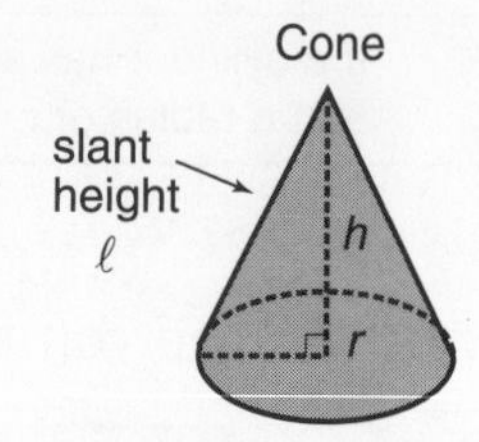

In the following formulas, L is lateral area, S is surface area, P is perimeter, and ℓ is slant height.

Regular Pyramids $L = \frac{1}{2}P\ell$

S = Lateral Area + Area of Base

Cones $L = \pi r\ell$

$S = \pi r\ell + \pi r^2$

Example: Find the surface area of the cone.

$S = \pi r\ell + \pi r^2$
$S = \pi(6)(10) + \pi(6)^2$
$S = 60\pi + 36\pi$
$S = 96\pi$ or about 301.6 cm

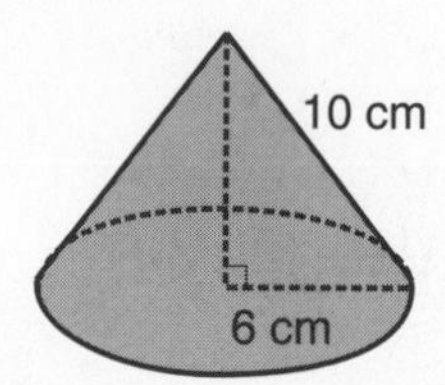

Find the lateral area and the surface area of each regular pyramid or cone. Round your answers to the nearest tenth.

1.

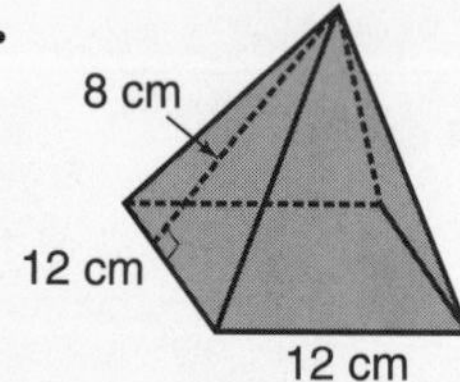

2.

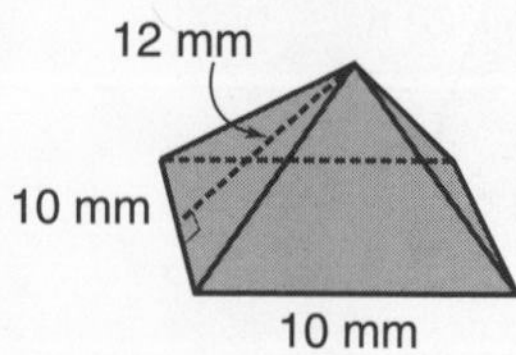

3.

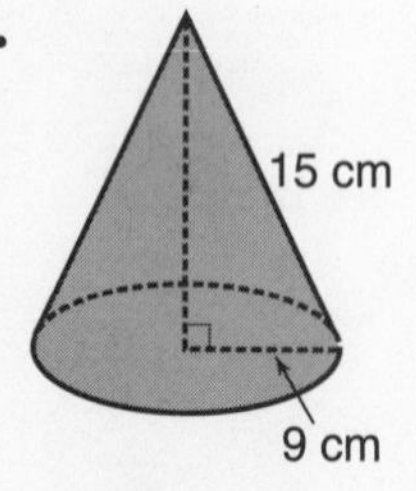

4.

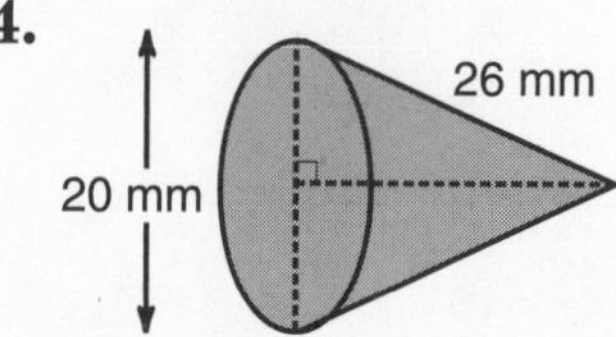

NAME ______________________ DATE __________ PERIOD ______

Study Guide

Student Edition
Pages 522–527

Volumes of Pyramids and Cones

Volume of a Cone	If a cone has a volume of V cubic units, a radius of r units, and a height of h units, then $V = \frac{1}{3}\pi r^2 h$.
Volume of a Pyramid	If a pyramid has a volume of V cubic units and a height of h units and the area of the base is B square units, then $V = \frac{1}{3}Bh$.

Examples: Find the volume of each solid.

1

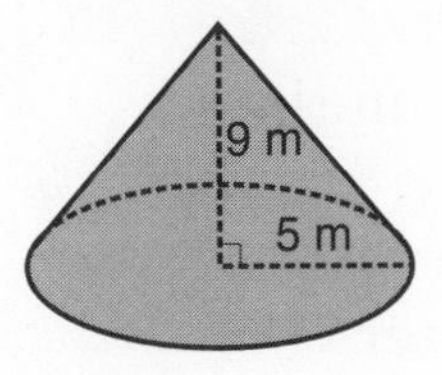

$V = \frac{1}{3}\pi r^2 h$

$V = \frac{1}{3}\pi (5^2)(9)$

$V = 75\pi$ or about 235.6 m^3

2

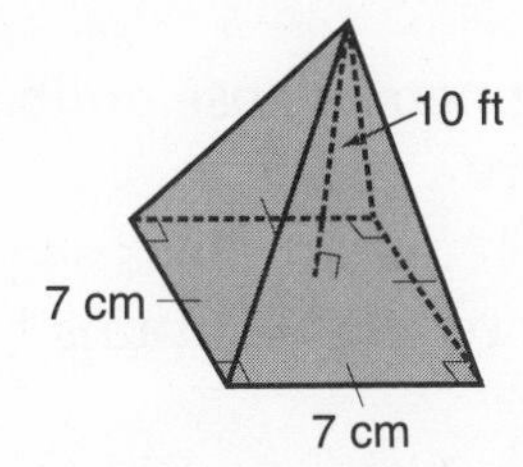

$V = \frac{1}{3}Bh$

$V = \frac{1}{3}(7 \cdot 7)\ 10$

$V = \frac{490}{3}$ or about 163.3 cm^3

Find the volume of each solid. Round your answers to the nearest tenth.

1.

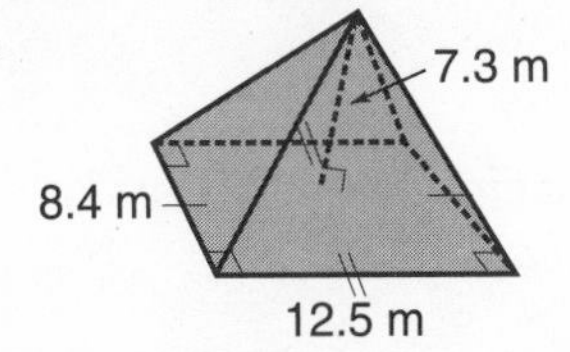

2.

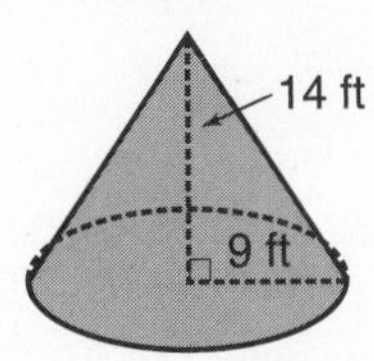

3.

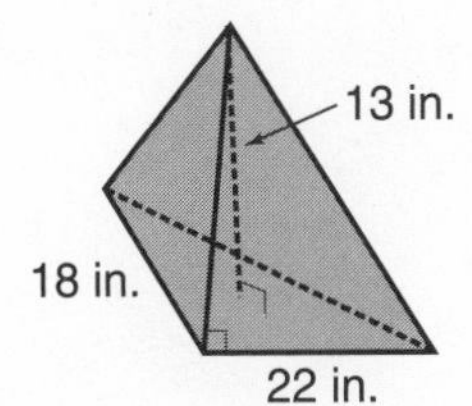

4.

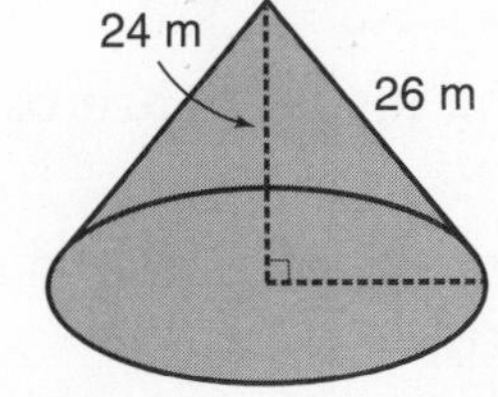

5.

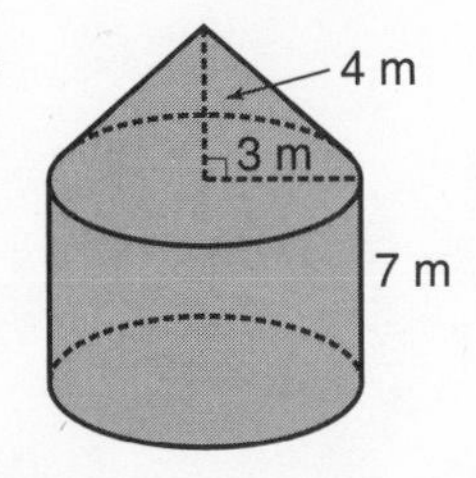

6.

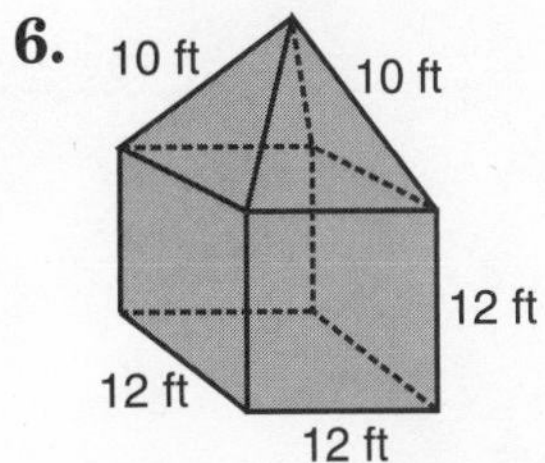

12-6

NAME ______________________ DATE __________ PERIOD ______

Study Guide

Student Edition
Pages 528–533

Spheres

The following is a list of definitions related to the study of spheres.

Sphere	the set of all points that are a given distance from a given point (center)
Radius	a segment whose endpoints are the center of the sphere and a point on the sphere
Chord	a segment whose endpoints are points on the sphere
Diameter	a chord that contains the sphere's center
Tangent	a line that intersects the sphere in exactly one point
Hemispheres	two congruent halves of a sphere separated by a great circle

***Describe each object as a model of a* circle, sphere, *or* neither.**

1. tennis ball can

2. pancake

3. sun

4. basketball rim

5. globe

6. lipstick container

***Determine whether each statement is* true *or* false.**

7. All lines intersecting a sphere are tangent to the sphere.

8. The eastern hemisphere of Earth is congruent to the western hemisphere of Earth.

NAME ______________________ DATE ________ PERIOD _____

Study Guide

Student Edition
Pages 534–539

Similarity of Solid Figures

Solids that have the same shape but are different in size are said to be **similar**. You can determine if two solids are similar by comparing the ratios **(scale factors)** of corresponding linear measurements. If the scale factor is 1:1, then the solids are **congruent**.

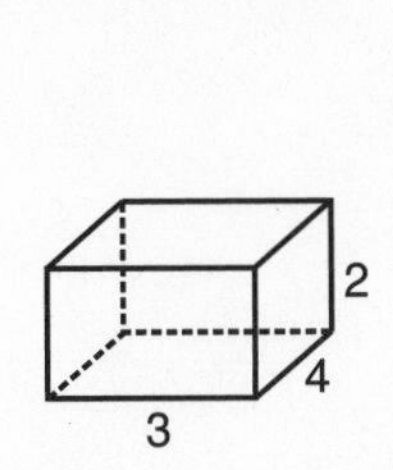

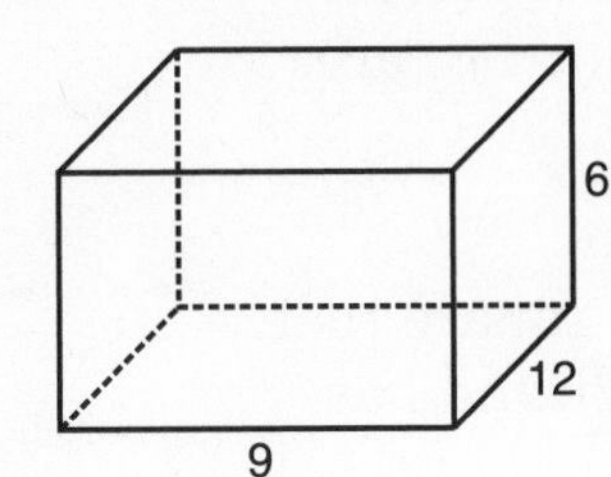

similar

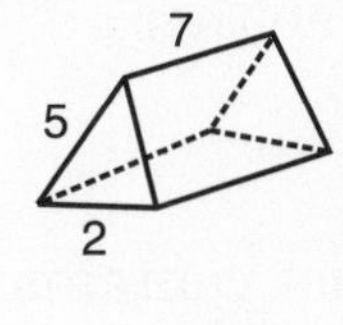

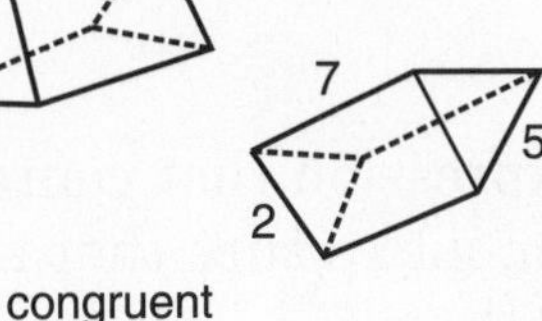

congruent

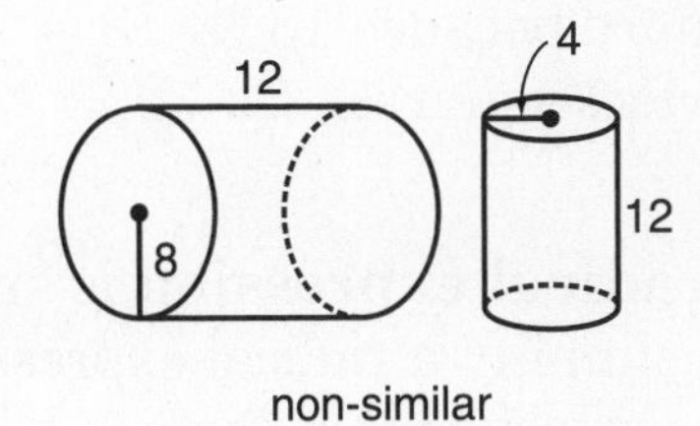

non-similar

***Determine if each pair of solids is* similar, congruent, *or* neither.**

1.

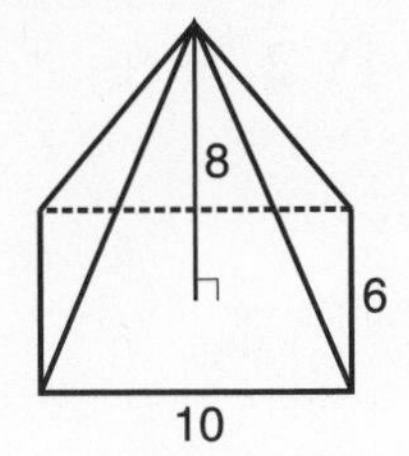

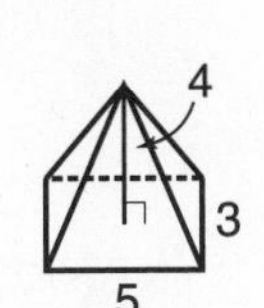

2.

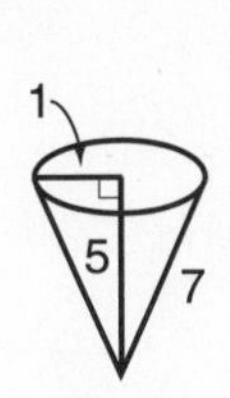

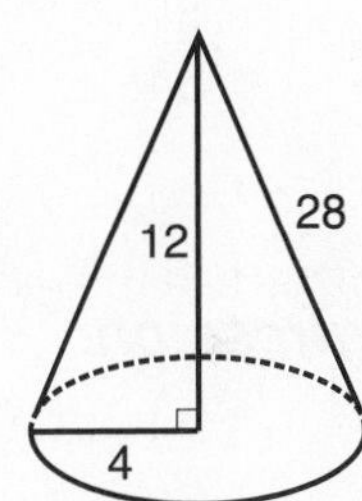

3.

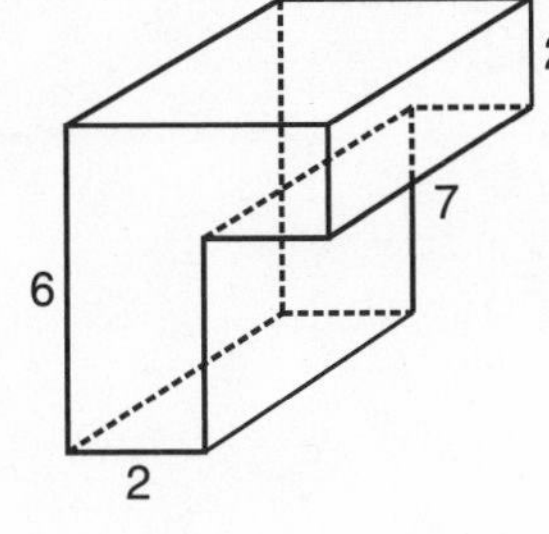

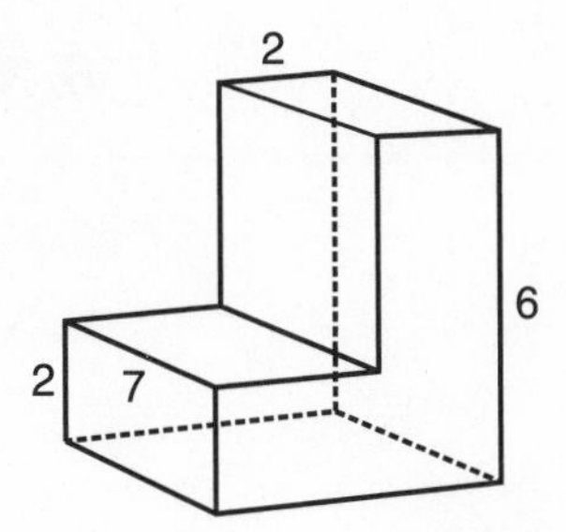

4.

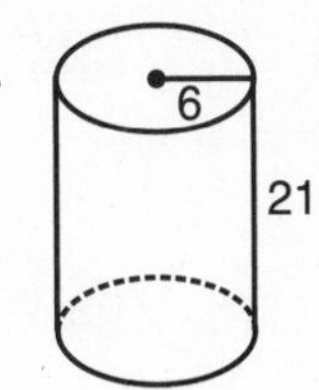

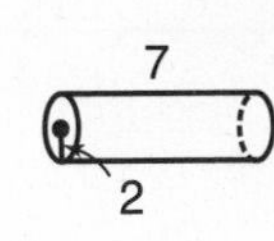

5.

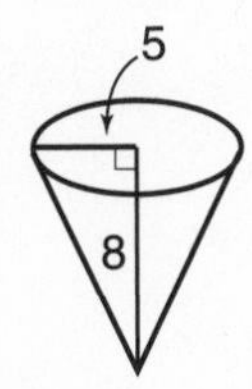

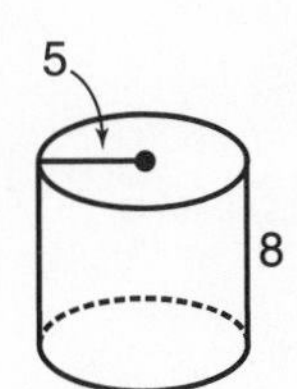

6.

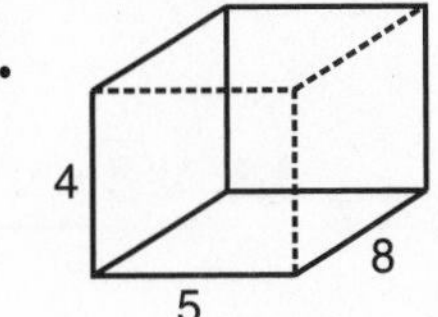

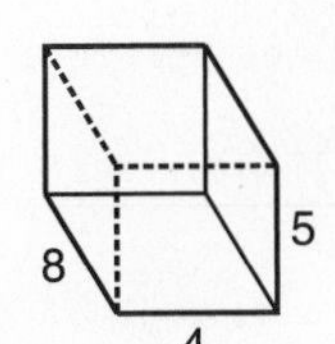

NAME ______________________ DATE __________ PERIOD ______

Study Guide

Student Edition
Pages 548–553

Simplifying Square Roots

Since $6 \times 6 = 36$, a **square root** of 36 is 6.

$$\sqrt{36} = 6$$

If the square root of a number is a whole number, the original number is called a **perfect square**. For example, 169 is a perfect square because $13 \times 13 = 169$. However, neither 168 nor 170 are perfect squares.

A **radical expression** is an expression that contains a square root. To simplify a radical expression, make sure that the radicand has no perfect square factors other than 1.

Examples: ***1*** Simplify $\sqrt{18}$.

$$\begin{aligned}\sqrt{18} &= \sqrt{3 \cdot 3 \cdot 2}\\ &= \sqrt{3 \cdot 3} \cdot \sqrt{2}\\ &= 3 \cdot \sqrt{2}\\ &= 3\sqrt{2}\end{aligned}$$

2 Simplify $\sqrt{4} \cdot \sqrt{8}$.

$$\begin{aligned}\sqrt{4} \cdot \sqrt{8} &= \sqrt{4 \cdot 8}\\ &= \sqrt{2 \cdot 2 \cdot 2 \cdot 2 \cdot 2}\\ &= \sqrt{2 \cdot 2} \cdot \sqrt{2 \cdot 2} \cdot \sqrt{2}\\ &= 2 \cdot 2 \cdot \sqrt{2} \text{ or } 4\sqrt{2}\end{aligned}$$

Simplify each expression.

1. $\sqrt{25}$ **2.** $\sqrt{64}$ **3.** $\sqrt{196}$

4. $\sqrt{900}$ **5.** $\sqrt{324}$ **6.** $\sqrt{529}$

7. $\sqrt{72}$ **8.** $\sqrt{24}$ **9.** $\sqrt{99}$

10. $\sqrt{300}$ **11.** $\sqrt{90}$ **12.** $\sqrt{75}$

13. $\sqrt{2} \cdot \sqrt{25}$ **14.** $\sqrt{3} \cdot \sqrt{32}$ **15.** $\sqrt{5} \cdot \sqrt{8}$

16. $\frac{\sqrt{4}}{\sqrt{9}}$ **17.** $\frac{\sqrt{16}}{\sqrt{36}}$ **18.** $\frac{\sqrt{2401}}{\sqrt{49}}$

NAME ____________________ DATE ________ PERIOD ______

Study Guide

45°-45°-90° Triangles

A special kind of right triangle is the 45°-45°-90° triangle. In a 45°-45°-90° triangle, the legs are of equal length and the hypotenuse is $\sqrt{2}$ times as long as a leg.

Example: Find the value of x.

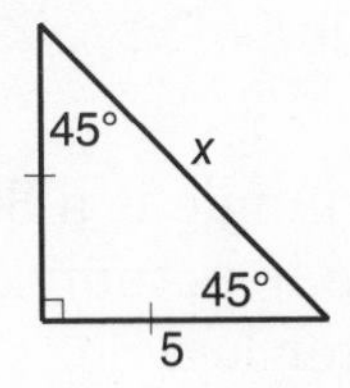

Since the triangle is a 45°–45°–90° triangle, the hypotenuse is $\sqrt{2}$ times as long as a leg. So, $x = 5\sqrt{2}$ or about 7.1.

Find the missing measures. Write all radicals in simplest form.

1.

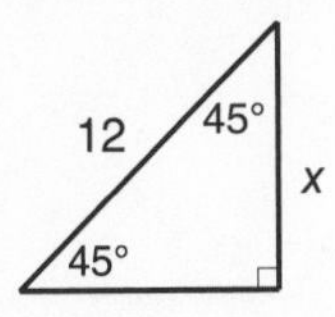

2.

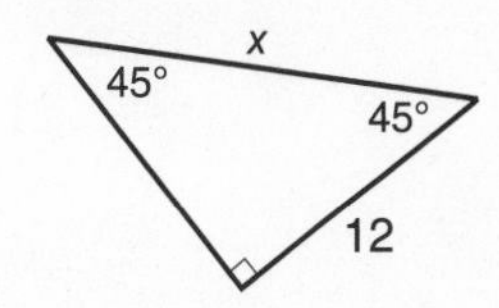

3.

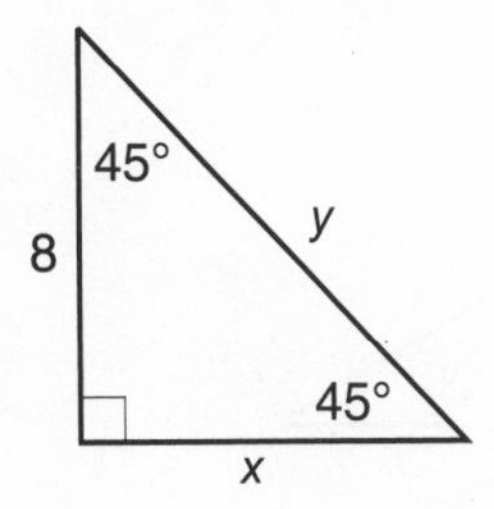

4.

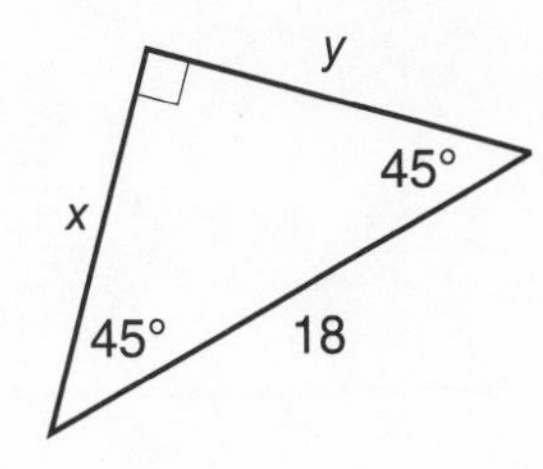

5.

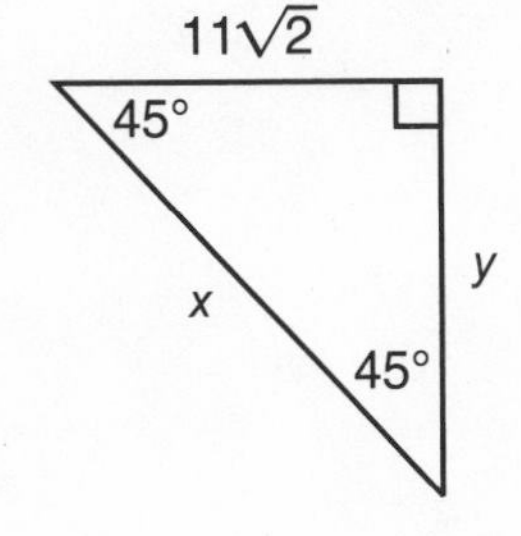

6.

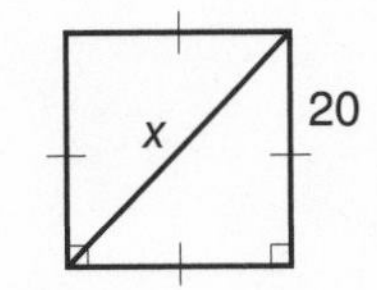

NAME ______________________ DATE __________ PERIOD _____

Study Guide

Student Edition
Pages 559–563

30°-60°-90° Triangles

Another special kind of right triangle is the 30°-60°-90° triangle. In a 30°-60°-90° triangle, the hypotenuse is twice as long as the shorter leg and the longer leg is $\sqrt{3}$ times as long as the shorter leg.

Example: Find the value of x.

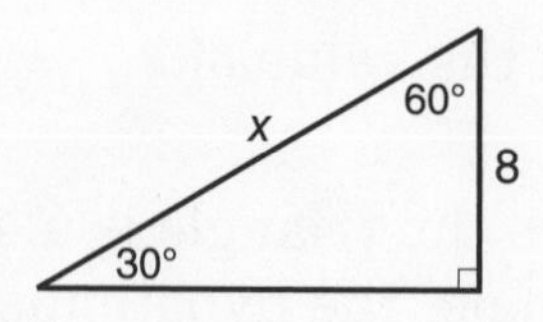

Since the triangle is a 30°-60°-90° triangle, the hypotenuse is twice as long as the shorter leg. So, $x = 2(8)$ or 16.

Find the missing measures. Write all radicals in simplest form.

1.

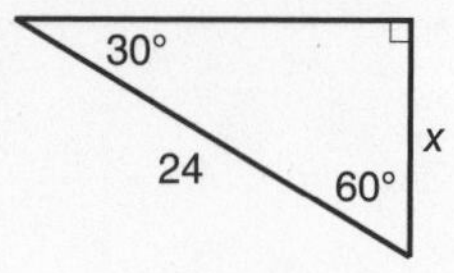

2.

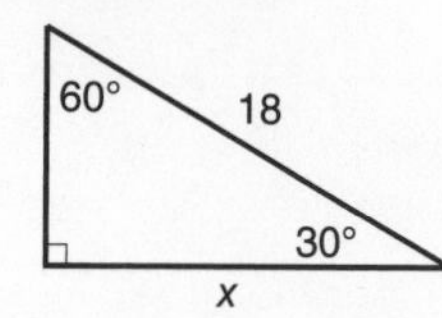

3.

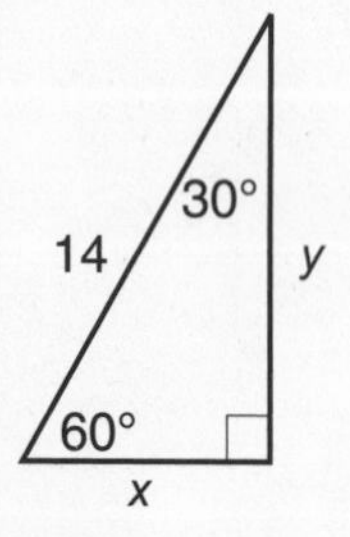

4.

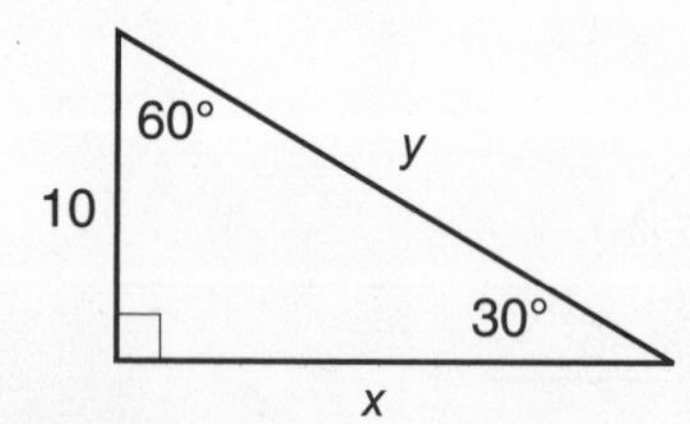

5.

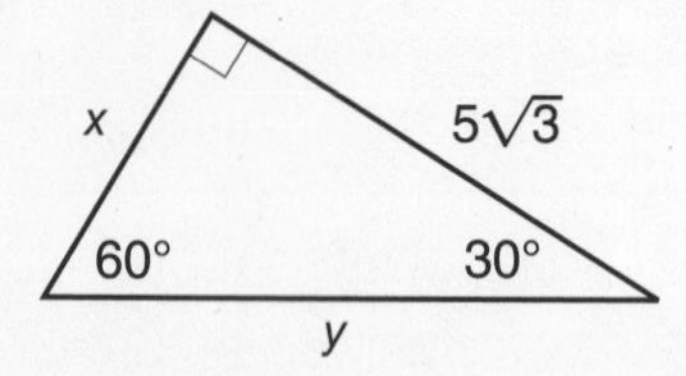

6.

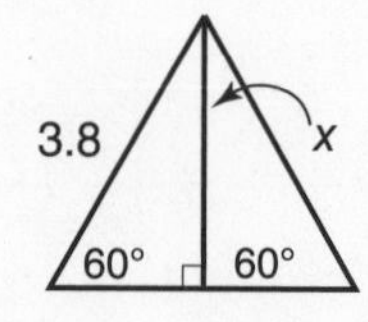

NAME ______________________ DATE __________ PERIOD ______

Study Guide

Tangent Ratio

Many problems in daily life can be solved by using trigonometry. Often such problems involve an **angle of elevation** or an **angle of depression**.

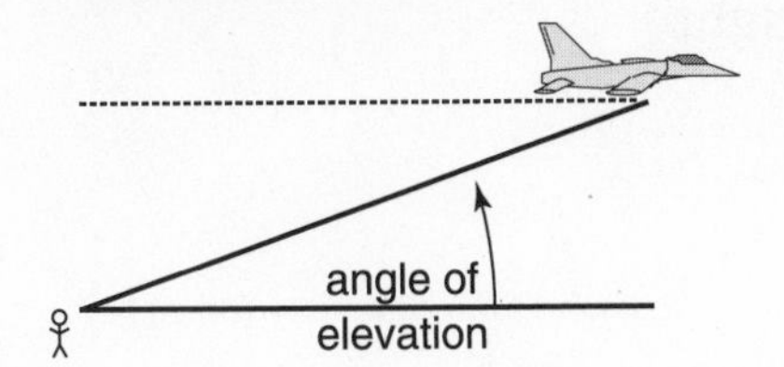

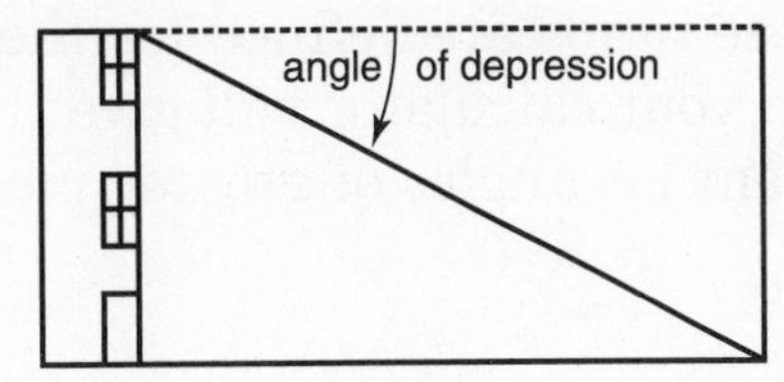

Example: The angle of elevation from point A to the top of a cliff is 38°. If point A is 80 feet from the base of the cliff, how high is the cliff?

Let x represent the height of the cliff.
Then $\tan 38° = \frac{x}{80}$.

$80 \tan 38° = x$

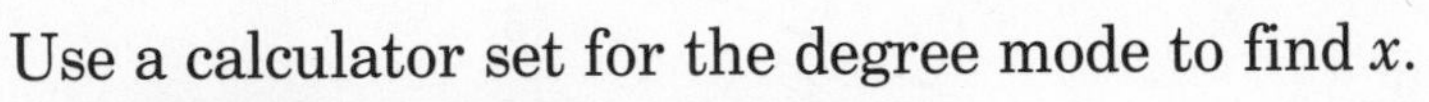

Use a calculator set for the degree mode to find x.

80 [×] [TAN] 38 [ENTER] 62.50285012

The cliff is about 63 feet high.

Solve each problem. Round to the nearest tenth.

1. From the top of a tower, the angle of depression to a stake on the ground is 72°. The top of the tower is 80 feet above ground. How far is the stake from the foot of the tower?

2. A tree 40 feet high casts a shadow 58 feet long. Find the measure of the angle of elevation of the sun.

3. A ladder leaning against a house makes an angle of 60° with the ground. The foot of the ladder is 7 feet from the foundation of the house. How far up the wall of the house does the ladder reach?

4. A balloon on a 40-foot string makes an angle of 50° with the ground. How high above the ground is the balloon if the hand of the person holding the balloon is 6 feet above the ground?

NAME ______________________ DATE __________ PERIOD _____

Study Guide

Sine and Cosine Ratios

A ratio of the lengths of two sides of a right triangle is called a **trigonometric ratio**. The three most common ratios are **sine**, **cosine**, and **tangent**. Their abbreviations are *sin*, *cos*, and *tan*, respectively. These ratios are defined for the acute angles of right triangles, though your calculator will give the values of sine, cosine, and tangent for angles of greater measure.

$\sin R = \dfrac{\text{leg opposite } \angle R}{\text{hypotenuse}} = \dfrac{r}{t}$

$\cos R = \dfrac{\text{leg adjacent to } \angle R}{\text{hypotenuse}} = \dfrac{s}{t}$

$\tan R = \dfrac{\text{leg opposite to } \angle R}{\text{leg adjacent to } \angle R} = \dfrac{r}{s}$

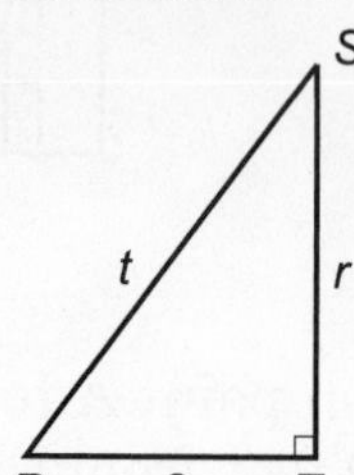

Example: Find sin *D*, cos *D*, and tan *D*. Express each ratio as a fraction and as a decimal rounded to the nearest thousandth.

$\sin D = \dfrac{5}{13} \approx 0.385$

$\cos D = \dfrac{12}{13} \approx 0.923$

$\tan D = \dfrac{5}{12} \approx 0.417$

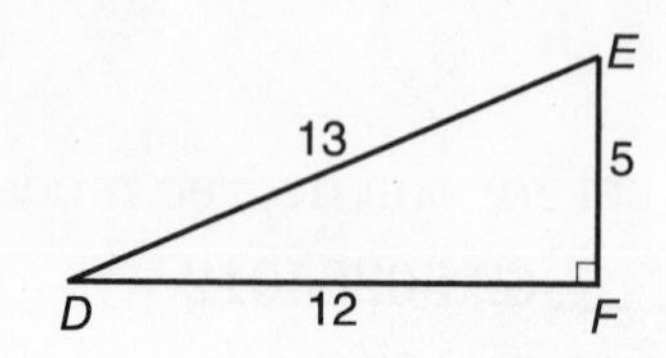

Find the indicated trigonometric ratio as a fraction and as a decimal rounded to the nearest ten-thousandth.

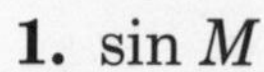

1. sin *M* **2.** cos *Z*

3. tan *L* **4.** sin *X*

5. cos *L* **6.** tan *Z*

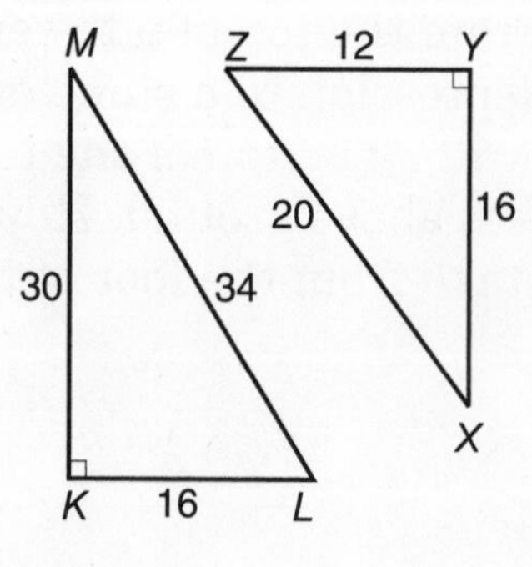

Find the value of each ratio to the nearest ten-thousandth.

7. sin 12° **8.** cos 32°

9. tan 74° **10.** sin 55°

14-1

NAME ______________________ DATE __________ PERIOD ______

Study Guide

Student Edition
Pages 586–591

Inscribed Angles

An **inscribed angle** of a circle is an angle whose vertex is on the circle and whose sides contain chords of the circle. We say that $\angle DEF$ *intercepts* $\widehat{DF}$. The following theorems involve inscribed angles.

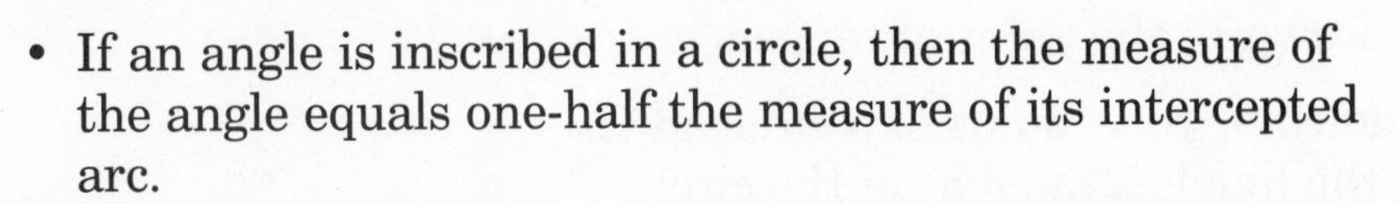

- If an angle is inscribed in a circle, then the measure of the angle equals one-half the measure of its intercepted arc.
- If inscribed angles of a circle or congruent circles intercept the same arc or congruent arcs, then the angles are congruent.
- If an inscribed angle of a circle intercepts a semicircle, then the angle is a right angle.

Example: In the circle above, find $m\angle DEF$ if $m\widehat{DF} = 28$.

Since $\angle DEF$ is an inscribed angle,
$m\angle DEF = \frac{1}{2} m\widehat{DF} = \frac{1}{2}(28)$ or 14.

In ⊙P, $\overline{RS} \parallel \overline{TV}$.

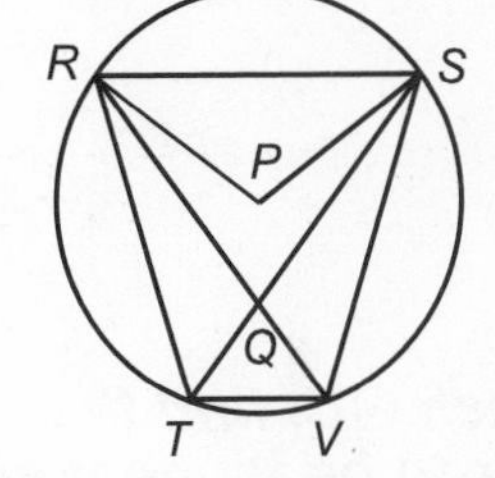

1. Name the intercepted arc for $\angle RTS$.

2. Name an inscribed angle.

3. Is $\angle RQS$ an inscribed angle?

In ⊙P, $m\widehat{SV} = 86$ and $m\angle RPS = 110$. Find each measure.

4. $m\angle PRS$ **5.** $m\widehat{RT}$ **6.** $m\angle RVT$ **7.** $m\angle SVT$

8. $m\angle TQV$ **9.** $m\angle RQT$ **10.** $m\angle QRT$ **11.** $m\widehat{RS}$

NAME ________________________ DATE ________ PERIOD ______

Study Guide

Student Edition
Pages 592–597

Tangents to a Circle

A **tangent** is a line in the plane of a circle that intersects the circle in exactly one point. Three important theorems involving tangents are the following.

- In a plane, if a line is a tangent to a circle, then it is perpendicular to the radius drawn to the point of tangency.
- In a plane, if a line is perpendicular to a radius of a circle at its endpoint on the circle, then the line is a tangent of the circle.
- If two segments from the same exterior point are tangent to a circle, then they are congruent.

Example: Find the value of x if $\overline{AB}$ is tangent to $\odot C$.

Tangent $\overline{AB}$ is perpendicular to radius $\overline{BC}$.
Also, $AC = AD + BC = 17$.

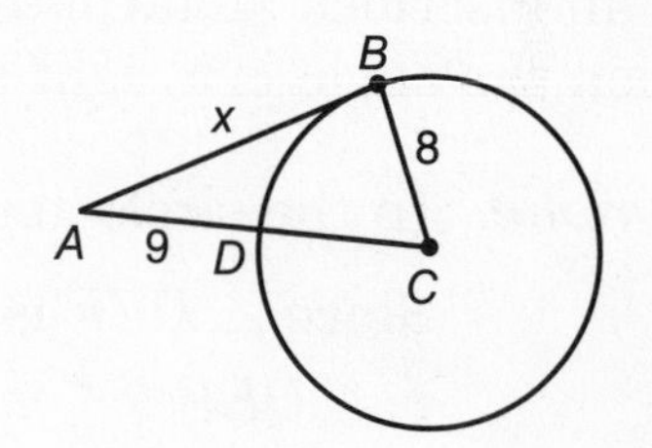

$$(AB)^2 + (BC)^2 = (AC)^2$$
$$x^2 + 8^2 = 17^2$$
$$x^2 + 64 = 289$$
$$x^2 = 225$$
$$x = 15$$

For each ⊙C, find the value of x. Assume segments that appear to be tangent are tangent.

1.

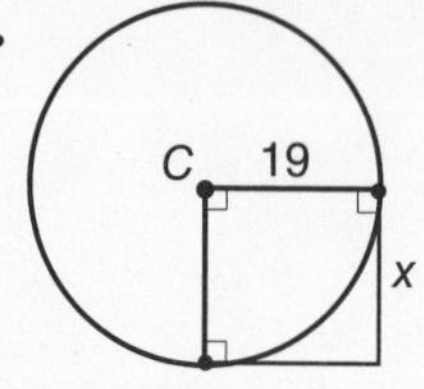

2.

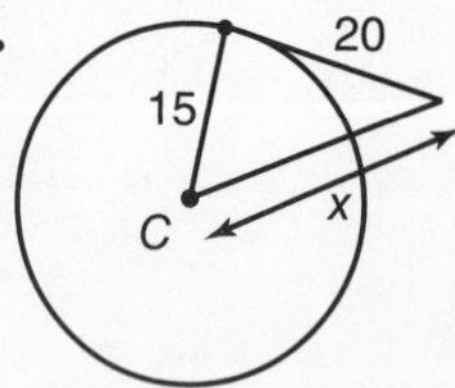

3.

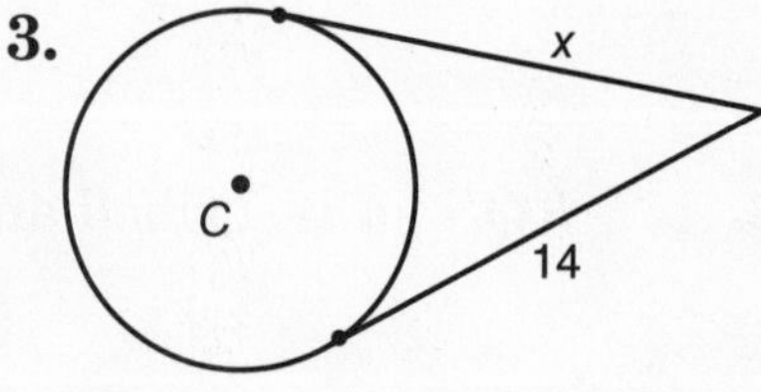

4.

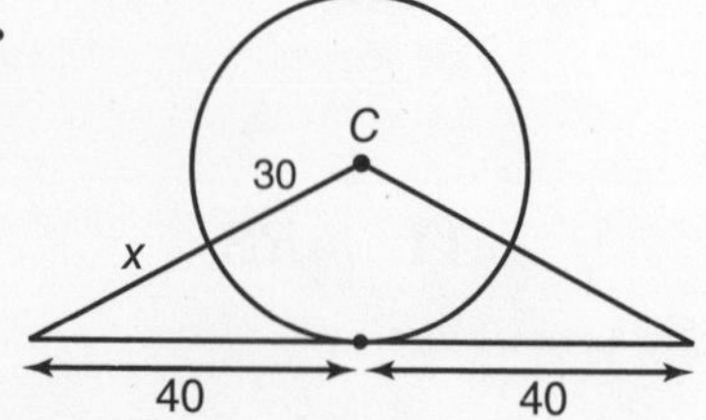

5.

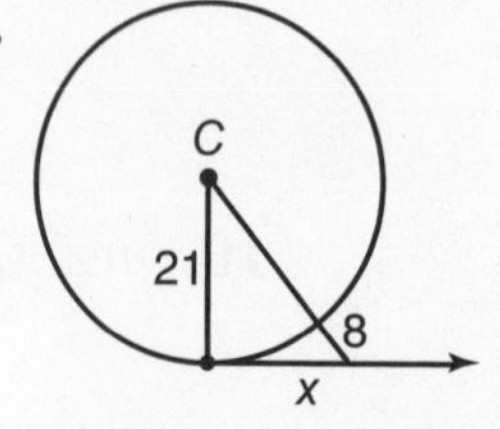

6.

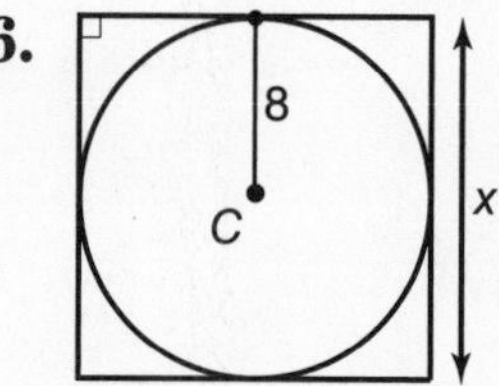

NAME ______________________ DATE ________ PERIOD ______

Study Guide

Student Edition
Pages 600–605

Secant Angles

A line that intersects a circle in exactly two points is called a **secant** of the circle. You can find the measures of angles formed by secants and tangents by using the following theorems.

- If a secant angle has its vertex inside a circle, then its degree measure is one-half the sum of the degree measures of the arcs intercepted by the angle and its vertical angle.
- If a secant angle has its vertex outside a circle, then its degree measure is one-half the difference of the degree measures of the intercepted arcs.

Example: Find the measure of $\angle MPN$.

You can use the last theorem above.

$$m\angle MPN = \frac{1}{2}(m\widehat{MN} - m\widehat{RS})$$
$$= \frac{1}{2}(34 - 18)$$
$$= \frac{1}{2}(16) \text{ or } 8$$

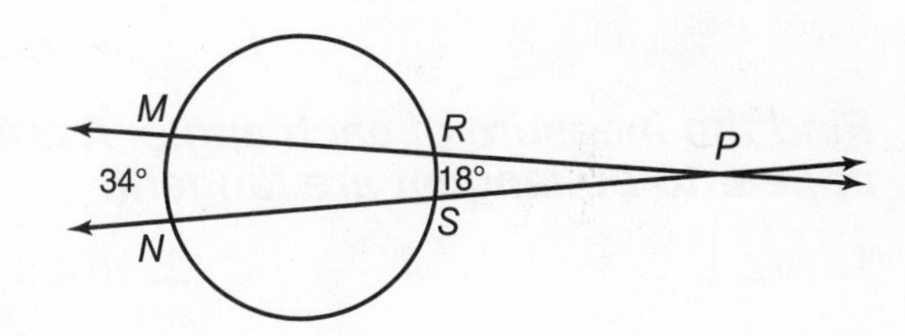

Find the measure of each numbered angle.

1.

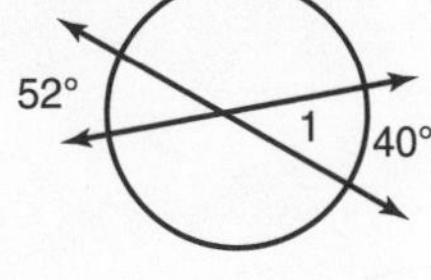

2.

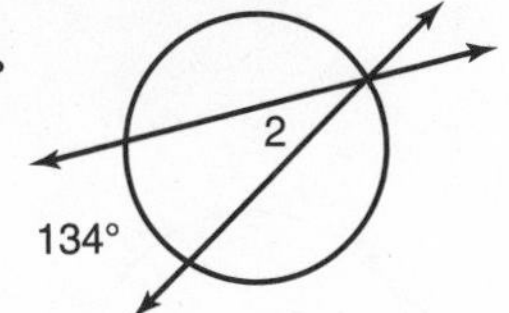

3.

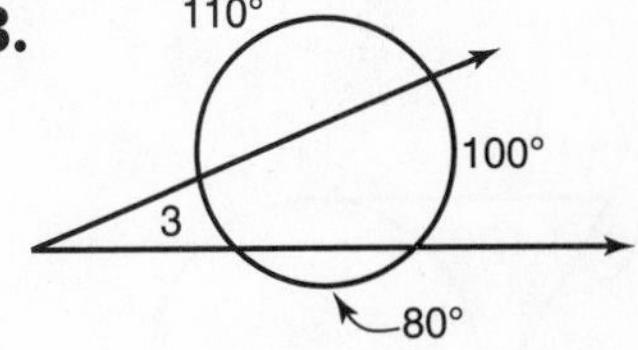

In each circle, find the value of x.

4.

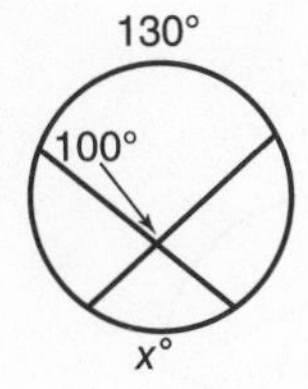

5.

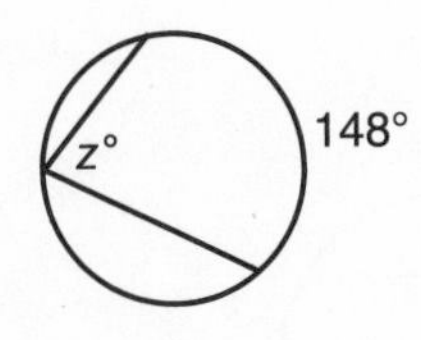

6.

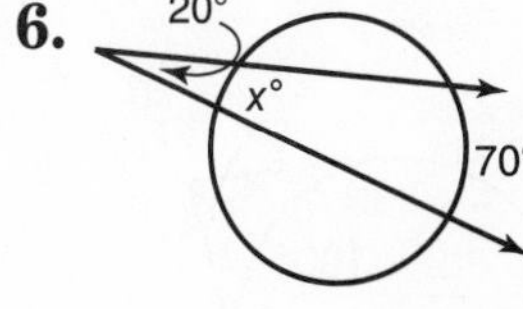

NAME ______________________________ DATE __________ PERIOD ______

14-4 Study Guide

Student Edition
Pages 606–611

Secant-Tangent Angles

You can find the measures of angles formed by secants and tangents by using the following theorems.

- If a secant-tangent angle has its vertex outside a circle, then its degree measure is one-half the difference of the degree measures of the intercepted arcs.
- If a secant-tangent angle has its vertex on a circle, then its degree measure is one-half the degree measure of the intercepted arc.
- The degree measure of a tangent-tangent angle is one-half the difference of the degree measures of the intercepted arcs.

Find the measure of each angle. Assume segments that appear to be tangent are tangent.

1. $m\angle 1$

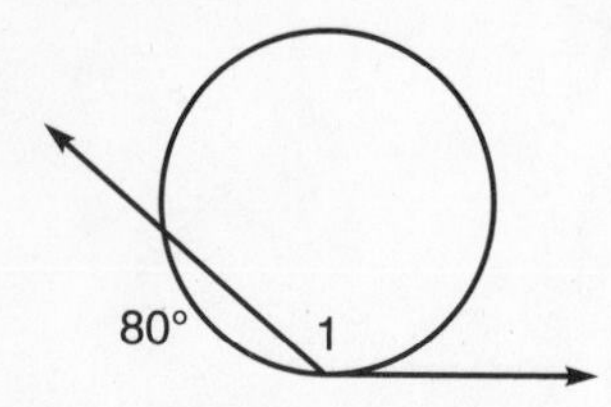

2. $m\angle 2$

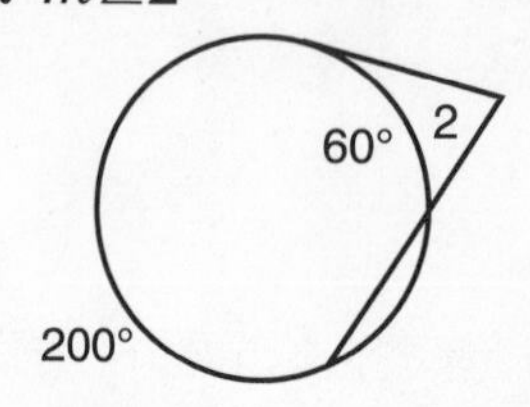

3. $m\angle 3$

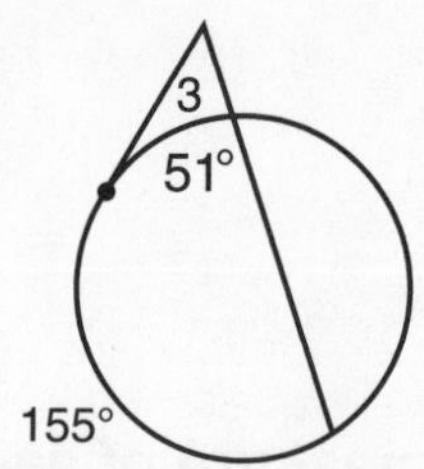

4. $m\angle 4$

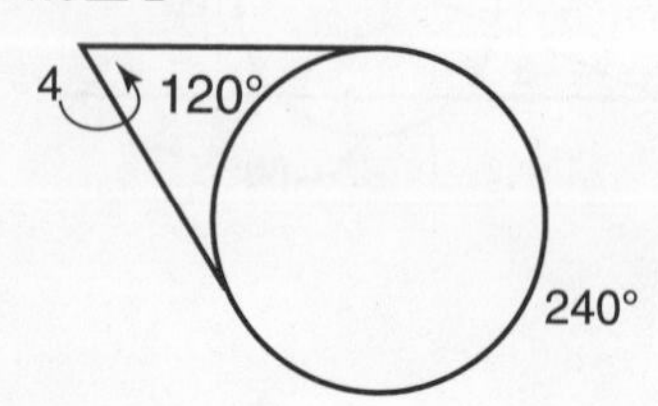

5. $m\angle 5$

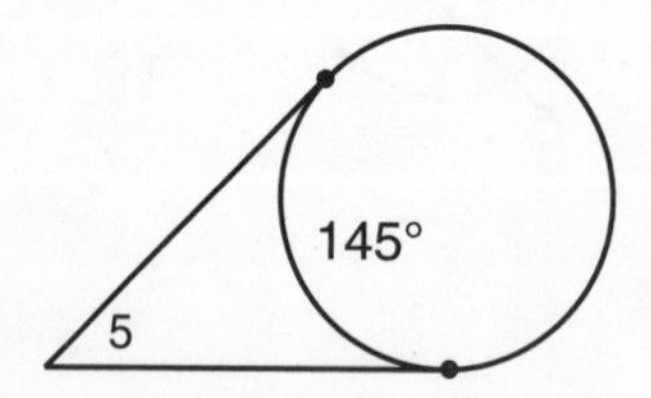

6. $m\angle 6$

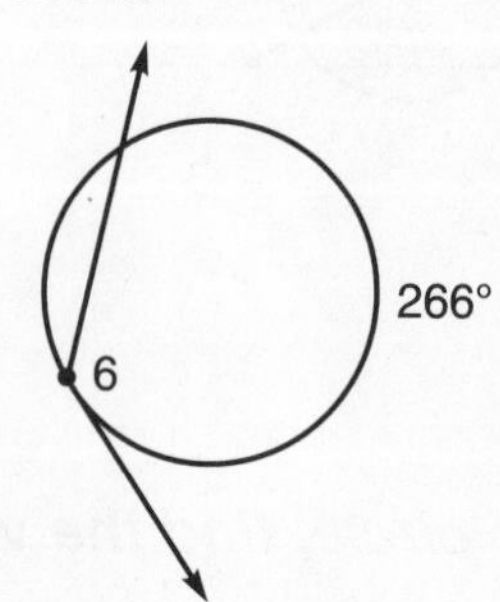

7. $m\angle 7$

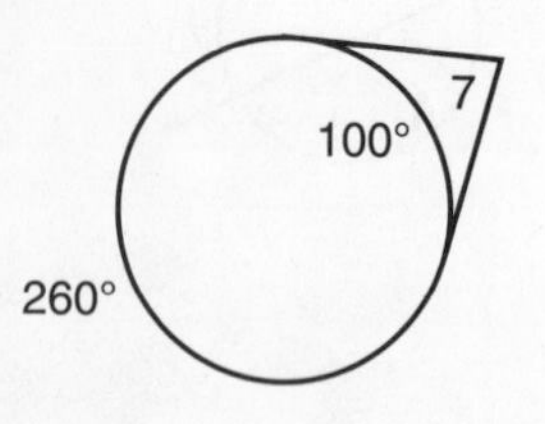

8. $m\angle 8$

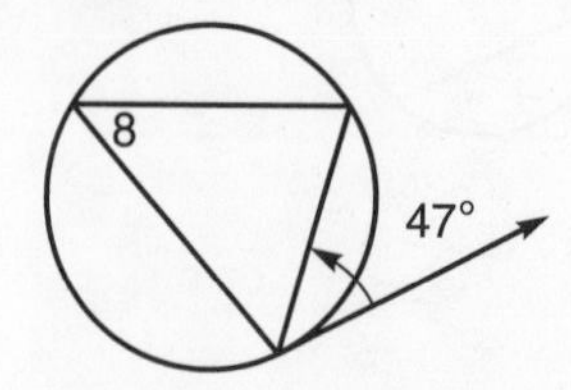

9. $m\angle 9$

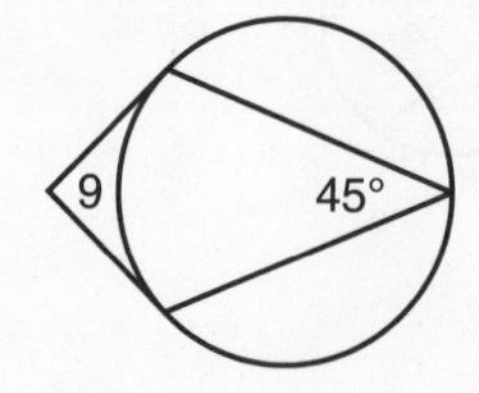

NAME ______________________ DATE ________ PERIOD ______

Study Guide

Student Edition
Pages 612–617

Segment Measures

The following theorems can be used to find the measure of special segments in a circle.

- If two chords of a circle intersect, then the products of the measures of the segments of the chords are equal.
- If two secant segments are drawn to a circle from an exterior point, then the product of the measures of one secant segment and its external secant segment is equal to the product of the measures of the other secant segment and its external secant segment.
- If a tangent segment and a secant segment are drawn to a circle from an exterior point, then the square of the measure of the tangent segment is equal to the product of the measures of the secant segment and its external secant segment.

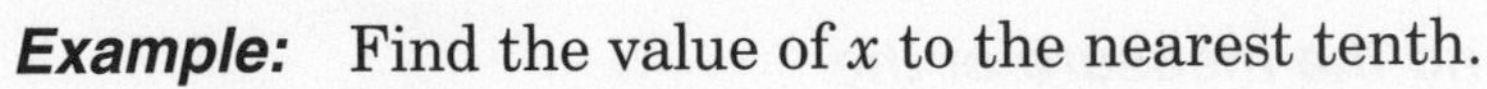

Example: Find the value of x to the nearest tenth.

$(AB)^2 = BD \cdot BC$ *Theorem 9–16*
$(18)^2 = (15 + x) \cdot 15$
$324 = 15x + 225$
$99 = 15x$
$6.6 = x$

Find the value of x to the nearest tenth. Assume segments that appear to be tangent are tangent.

1.
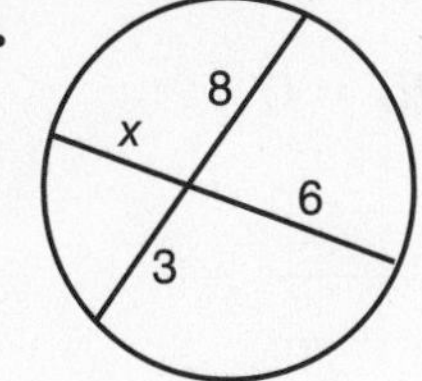

2.
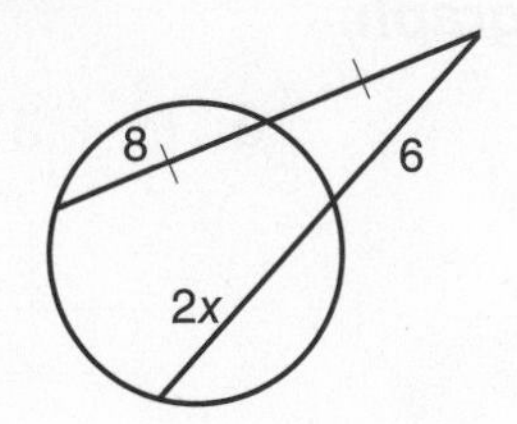

3.
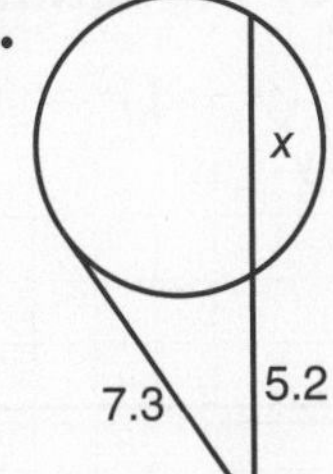

4.
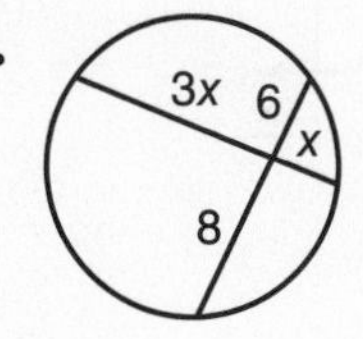

5.
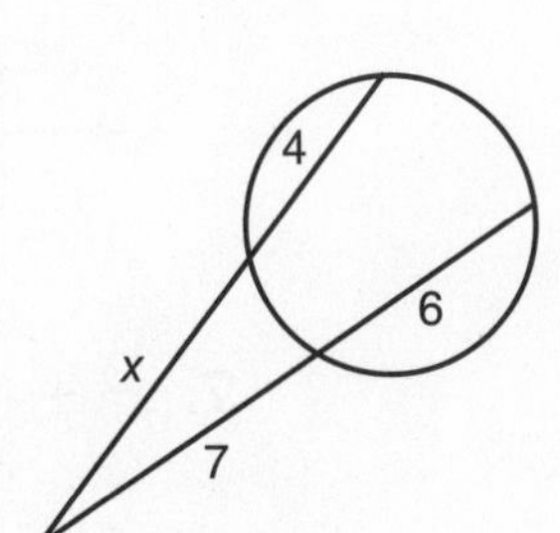

6.
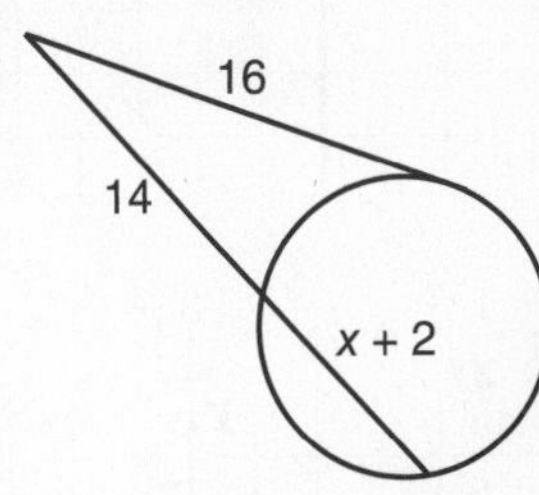

7.
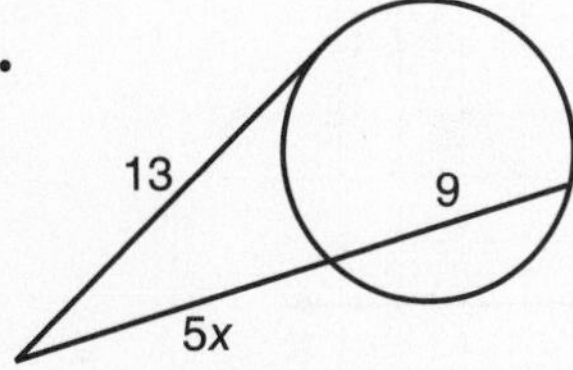

8.
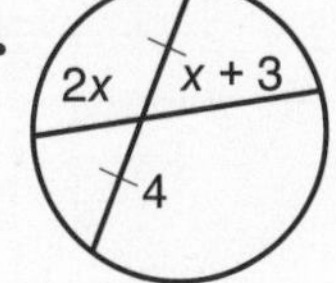

9.
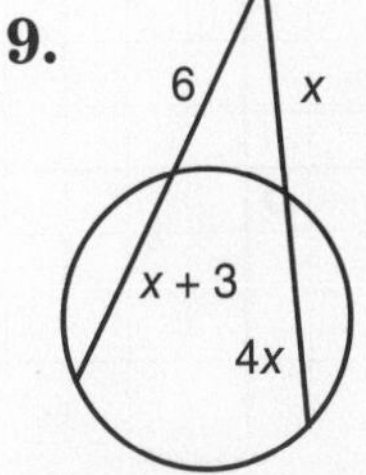

NAME ______________________ DATE __________ PERIOD ______

Study Guide

Student Edition
Pages 618–623

Equations of Circles

The general equation for a circle is derived from using the distance formula given the coordinates of the center of the circle and the measure of its radius. An equation for a circle with center (h, k) and a radius of r units is $(x - h)^2 + (y - k)^2 = r^2$.

Example: Graph the circle whose equation is $(x + 3)^2 + (y - 1)^2 = 16$.

$(x - h)^2 + (y - k)^2 = r^2$ *General Equation*

$(x - (-3))^2 + (y - 1)^2 = (\sqrt{16})^2$ *Substitution*

Therefore, $h = -3$, $k = 1$, and $r = \sqrt{16} = 4$.
The center is at (−3, 1) and the radius is 4 units.

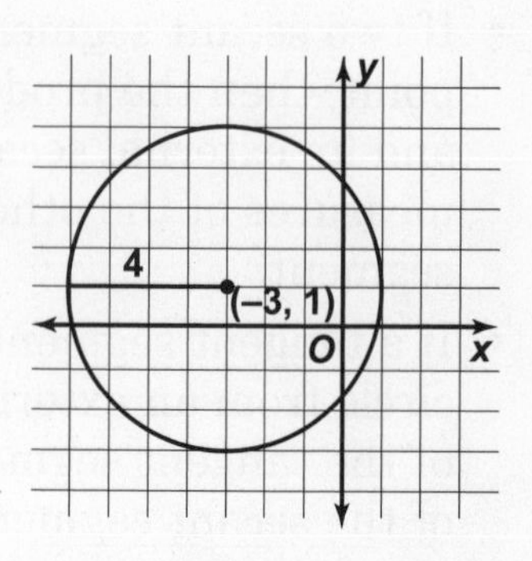

Find the coordinates of the center and the measure of the radius for each circle whose equation is given.

1. $(x - 7.2)^2 + (y + 3.4)^2 = 14.44$

2. $\left(x + \frac{1}{2}\right)^2 + (y - 2)^2 = \frac{16}{25}$

3. $(x - 6)^2 + (y - 3)^2 - 25 = 0$

Graph each circle whose equation is given. Label the center and measure of the radius on each graph.

4. $(x - 2.5)^2 + (y + 1)^2 = 12.25$

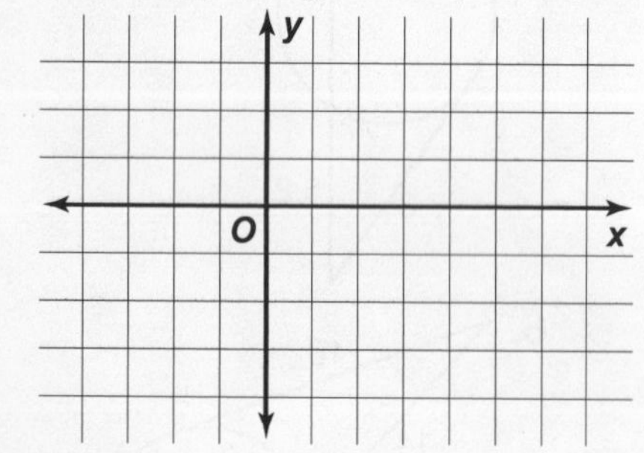

5. $(x + 3)^2 + (y - 4)^2 - 2.25 = 0$

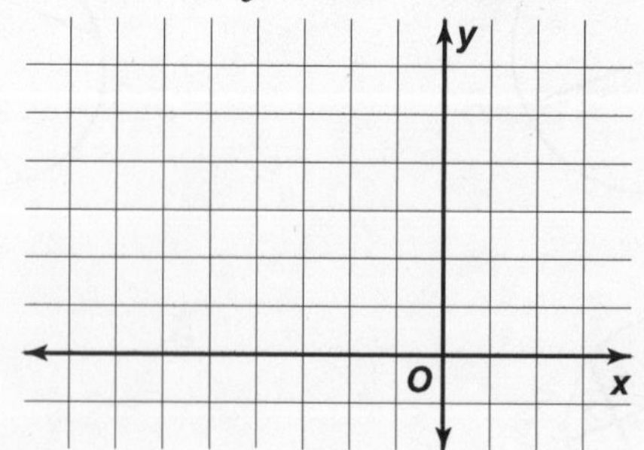

6. $\left(x - \frac{1}{2}\right)^2 + \left(y - \frac{3}{4}\right)^2 = 1$

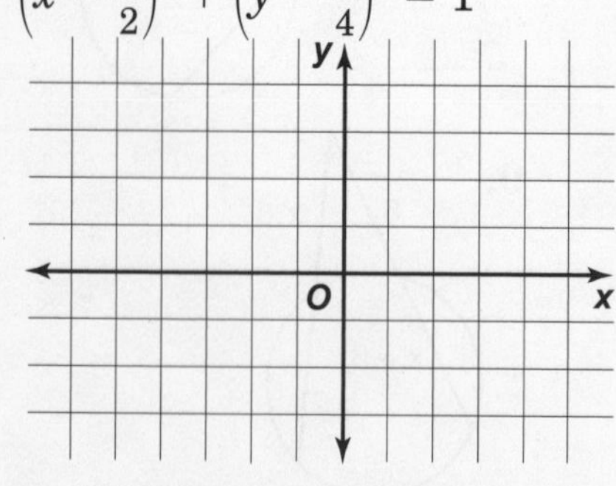

7. $x^2 + (y - 2)^2 = 9$

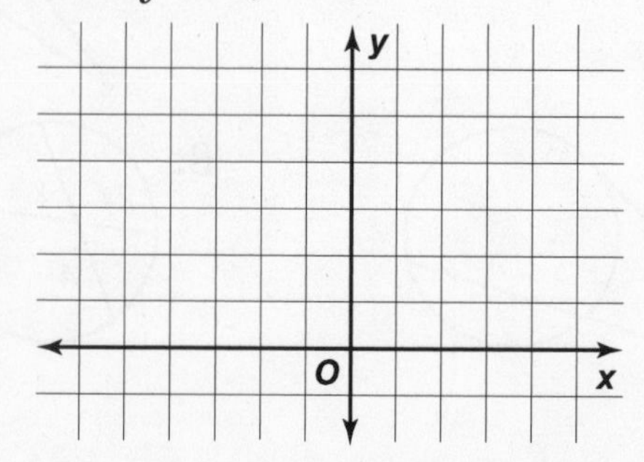

NAME ______________________ DATE __________ PERIOD ______

Study Guide

Student Edition
Pages 632–637

Logic and Truth Tables

A **statement** is any sentence that is either true or false, but not both. The table below lists different kinds of statements.

Term	Symbol	Definition
negation	$\sim p$	not p
conjunction	$p \wedge q$	p and q
disjunction	$p \vee q$	p or q
conditional	$p \rightarrow q$	if p, then q
converse	$q \rightarrow p$	if q, then p

Every statement has a **truth value**. It is convenient to organize the truth values in a **truth table** like the one shown at the right.

Conjunction		
p	q	$p \wedge q$
T	T	T
T	F	F
F	T	F
F	F	F

Complete a truth table for each compound statement.

1. $p \vee q$

p	q	$p \vee q$

2. $\sim p \vee q$

p	$\sim p$	q	$\sim p \vee q$

3. $p \wedge \sim q$

p	q	$\sim q$	$p \wedge \sim q$

4. $\sim(p \wedge q)$

p	q	$p \wedge q$	$\sim(p \wedge q)$

Study Guide

Student Edition
Pages 638–643

Deductive Reasoning

Two important laws used frequently in deductive reasoning are the **Law of Detachment** and the **Law of Syllogism**. In both cases you reach conclusions based on if-then statements.

Law of Detachment	Law of Syllogism
If $p \to q$ is a true conditional and p is true, then q is true.	If $p \to q$ and $q \to r$ are true conditionals, then $p \to r$ is also true.

Example: Determine if statement (3) follows from statements (1) and (2) by the Law of Detachment or the Law of Syllogism. If it does, state which law was used.

(1) If you break an item in a store, you must pay for it.
(2) Jill broke a vase in Potter's Gift Shop.
(3) Jill must pay for the vase.

Yes, statement (3) follows from statements (1) and (2) by the Law of Detachment.

***Determine if a valid conclusion can be reached from the two true statements using the Law of Detachment or the Law of Syllogism. If a valid conclusion is possible, state it and the law that is used. If a valid conclusion does not follow, write* no valid conclusion.**

1. (1) If a number is a whole number, then it is an integer.
(2) If a number is an integer, then it is a rational number.

2. (1) If a dog eats Dogfood Delights, the dog is happy.
(2) Fido is a happy dog.

3. (1) If people live in Manhattan, then they live in New York.
(2) If people live in New York, then they live in the United States.

4. (1) Angles that are complementary have measures with a sum of 90.

(2) $\angle A$ and $\angle B$ are complementary.

5. (1) All fish can swim.
(2) Fonzo can swim.

6. Look for a Pattern Find the next number in the list 83, 77, 71, 65, 59 and make a conjecture about the pattern.

NAME ______________________ DATE ________ PERIOD ______

15-3 Study Guide

Student Edition
Pages 644–648

Paragraph Proofs

A **proof** is a logical argument in which each statement you make is backed up by a reason that is accepted as true. In a **paragraph proof**, you write your statements and reasons in paragraph form.

Example: Write a paragraph proof for the conjecture.

Given: $\square WXYZ$

Prove: $\angle W$ and $\angle X$ are supplementary.
$\angle X$ and $\angle Y$ are supplementary.
$\angle Y$ and $\angle Z$ are supplementary.
$\angle Z$ and $\angle W$ are supplementary.

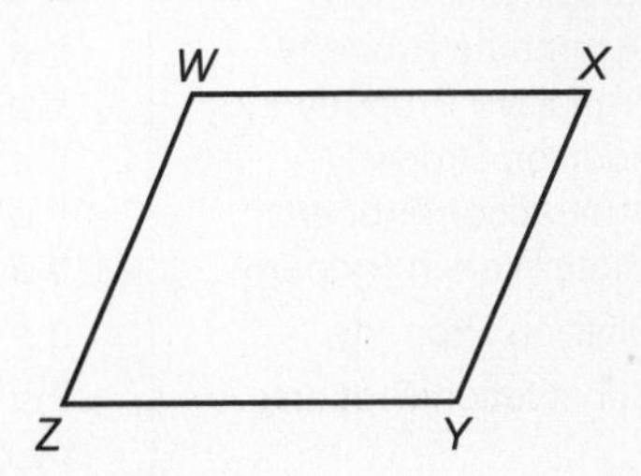

By the definition of a parallelogram, $\overline{WX} \parallel \overline{ZY}$ and $\overline{WZ} \parallel \overline{XY}$. For parallels $\overline{WX}$ and $\overline{ZY}$, $\overline{WZ}$ and $\overline{XY}$ are transversals; for parallels $\overline{WZ}$ and $\overline{XY}$, $\overline{WX}$ and $\overline{ZY}$ are transversals. Thus, the consecutive interior angles on the same side of a transversal are supplementary. Therefore, $\angle W$ and $\angle X$, $\angle X$ and $\angle Y$, $\angle Y$ and $\angle Z$, $\angle Z$ and $\angle W$ are supplementary.

Write a paragraph proof for each conjecture.

1. Given: $\angle PSU \cong \angle PTR$
$\overline{SU} \cong \overline{TR}$

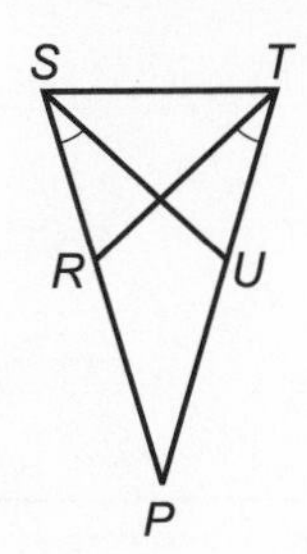

Prove: $\overline{SP} \cong \overline{TP}$

2. Given: $\triangle DEF$ and $\triangle RST$ are rt. triangles. $\angle E$ and $\angle S$ are right angles. $\overline{EF} \cong \overline{ST}$ and $\overline{ED} \cong \overline{SR}$

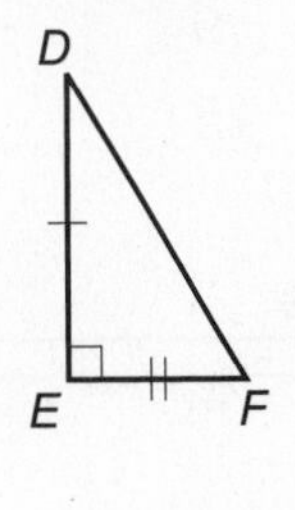

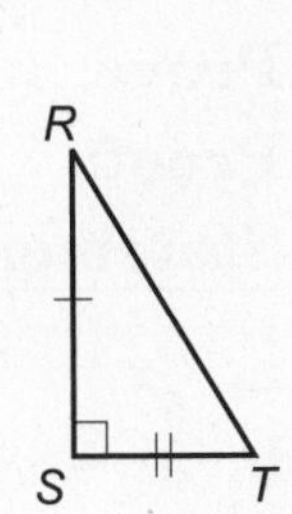

Prove: $\triangle DEF \cong \triangle RST$

15-4

NAME ______________________ DATE __________ PERIOD ______

Study Guide

Student Edition
Pages 649–653

Preparing for Two-Column Proofs

Many rules from algebra are used in geometry.

Properties of Equality for Real Numbers	
Reflexive Property	$a = a$
Symmetric Property	If $a = b$, then $b = a$.
Transitive Property	If $a = b$ and $b = c$, then $a = c$.
Addition Property	If $a = b$, then $a + c = b + c$.
Subtraction Property	If $a = b$, then $a - c = b - c$.
Multiplication Property	If $a = b$, then $a \cdot c = b \cdot c$.
Division Property	If $a = b$ and $c \neq 0$, then $\frac{a}{c} = \frac{b}{c}$.
Substitution Property	If $a = b$, then a may be replaced by b in any equation or expression.
Distributive Property	$a(b + c) = ab + ac$

Example: Prove that if $4x - 8 = -8$, then $x = 0$.

Given: $4x - 8 = -8$

Prove: $x = 0$

Proof:

Statements	Reasons
a. $4x - 8 = -8$	**a.** Given
b. $4x = 0$	**b.** Addition Property (=)
c. $x = 0$	**c.** Division Property (=)

Name the property that justifies each statement.

1. Prove that if $\frac{3}{5}x = -9$, then $x = -15$.

Given: $\frac{3}{5}x = -9$

Prove: $x = -15$

Proof:

Statements	Reasons
a. $\frac{3}{5}x = -9$	**a.** ______________
b. $3x = -45$	**b.** ______________
c. $x = -15$	**c.** ______________

2. Prove that if $3x - 2 = x - 8$, then $x = -3$.

Given: $3x - 2 = x - 8$

Prove: $x = -3$

Proof:

Statements	Reasons
a. $3x - 2 = x - 8$	**a.** ______________
b. $2x - 2 = -8$	**b.** ______________
c. $2x = -6$	**c.** ______________
d. $x = -3$	**d.** ______________

15-5

NAME ______________________ DATE ________ PERIOD ______

Study Guide

Student Edition
Pages 654-659

Two Column Proofs

The reasons necessary to complete the following proof are scrambled up below. To complete the proof, number the reasons to match the corresponding statements.

Given: $\overline{CD} \perp \overline{BE}$
$\overline{AB} \perp \overline{BE}$
$\overline{AD} \cong \overline{CE}$
$\overline{BD} \cong \overline{DE}$

Prove: $\overline{AD} \parallel \overline{CE}$

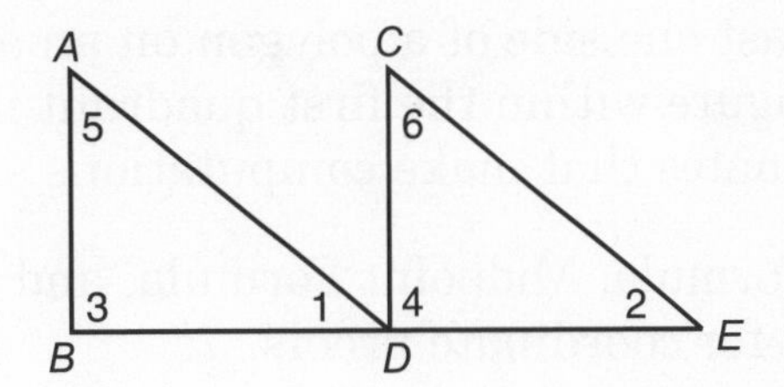

Proof:

Statements	Reasons
1. $\overline{CD} \perp \overline{BE}$	**1.** Definition of Right Triangle
2. $\overline{AB} \perp \overline{BE}$	**2.** Given
3. $\angle 3$ and $\angle 4$ are right angles.	**3.** Given
4. $\triangle ABD$ and $\triangle CDE$ are right triangles.	**4.** Definition of Perpendicular Lines
5. $\overline{AD} \cong \overline{CE}$	**5.** Given
6. $\overline{BD} \cong \overline{DE}$	**6.** CPCTC
7. $\triangle ABD \cong \triangle CDE$	**7.** In a plane, if two lines are cut by a transcersal so that a pair of corresponding angles is congruent, then the lines are parallel. (Postulate 4-2)
8. $\angle 1 \cong \angle 2$	**8.** Given
9. $\overline{AD} \parallel \overline{CE}$	**9.** HL

15-6

NAME ______________________ DATE ________ PERIOD _____

Study Guide

Student Edition
Pages 660–665

Coordinate Proofs

You can place figures in the coordinate plane and use algebra to prove theorems. The following guidelines for positioning figures can help keep the algebra simple.

- Use the origin as a vertex or center.
- Place at least one side of a polygon on an axis.
- Keep the figure within the first quadrant if possible.
- Use coordinates that make computations simple.

The Distance Formula, Midpoint Formula, and Slope Formula are useful tools for coordinate proofs.

Example: Use a coordinate proof to prove that the diagonals of a rectangle are congruent.

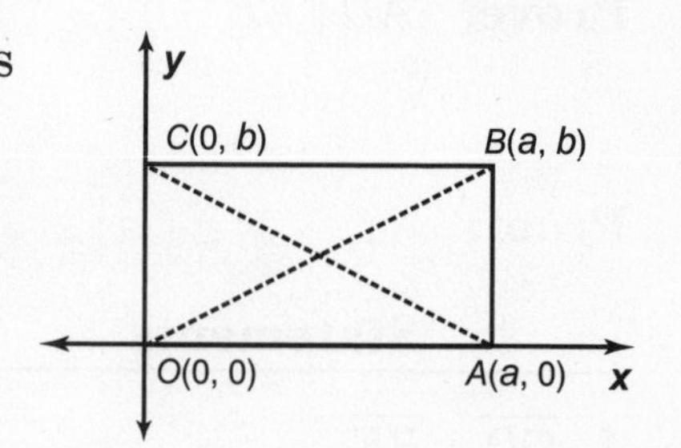

Use (0, 0) as one vertex. Place another vertex on the x-axis at $(a, 0)$ and another on the y-axis at $(0, b)$. The fourth vertex must then be (a, b).

Use the Distance Formula to find OB and AC.

$$OB = \sqrt{(a - 0)^2 + (b - 0)^2} = \sqrt{a^2 + b^2}$$
$$AC = \sqrt{(0 - a)^2 + (b - 0)^2} = \sqrt{a^2 + b^2}$$

Since $OB = AC$, the diagonals are congruent.

Name the missing coordinates in terms of the given variables.

1. *ABCD* is a rectangle.

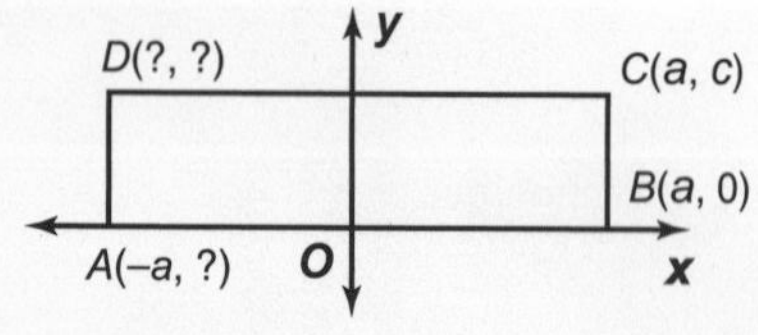

2. *HIJK* is a parallelogram.

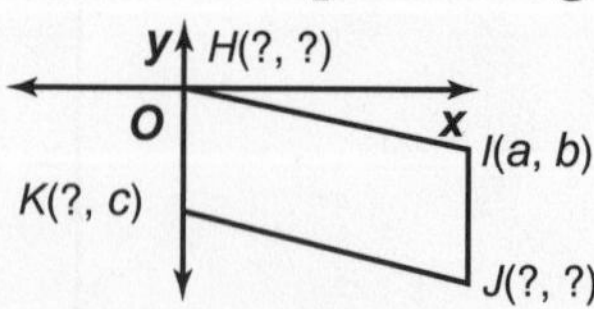

3. Use a coordinate proof to show that the opposite sides of any parallelogram are congruent.

NAME ______________________ DATE ________ PERIOD ______

Study Guide

Student Edition
Pages 676–680

Solving Systems of Equations by Graphing

The equations $y = -\frac{1}{2}x + 2$ and $y = 3x - 5$ together are called a **system of equations**.

The **solution** to this system is the ordered pair that is the solution of both equations. To solve a system of equations, graph each equation on the same coordinate plane. The point where both graphs intersect is the solution of the system of equations.

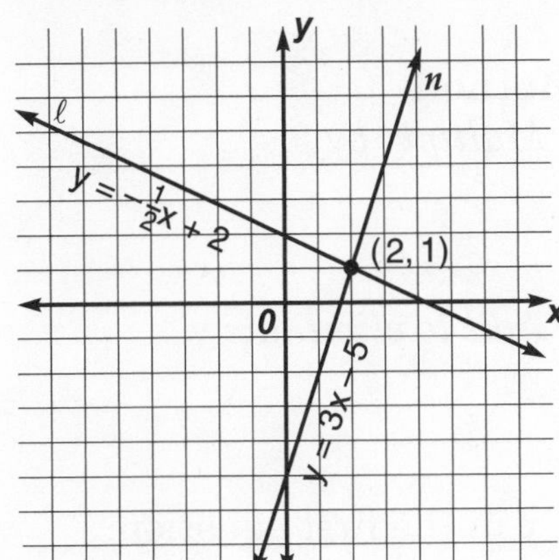

Line ℓ is the graph of $y = -\frac{1}{2}x + 2$.

Line n is the graph of $y = 3x - 5$.

The lines intersect at (2, 1). Therefore, the solution to the system of equations is (2, 1).

Solve each system of equations by graphing.

1. $y = x - 3$
$y = -3x + 1$

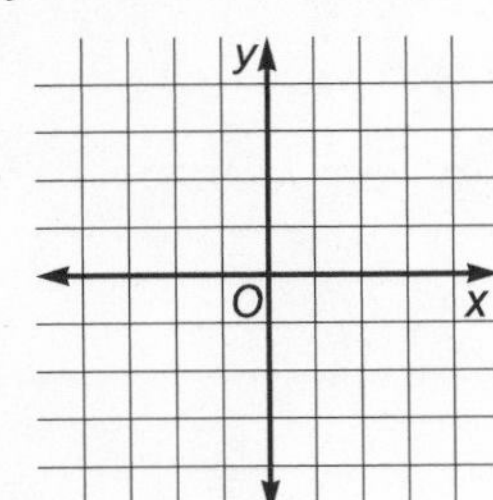

2. $y = 2x$
$y = x + 1$

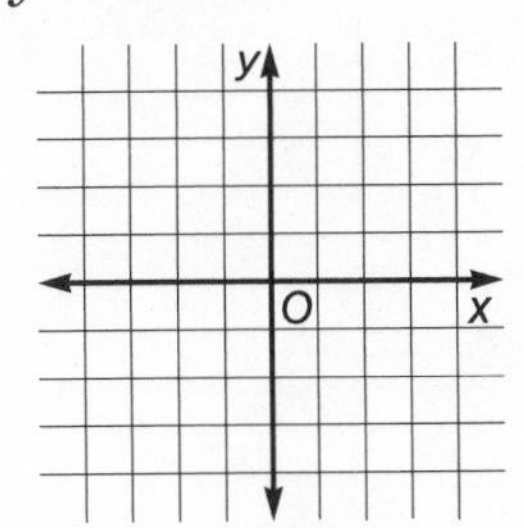

3. $y = 4x + 5$
$y = 4x - 1$

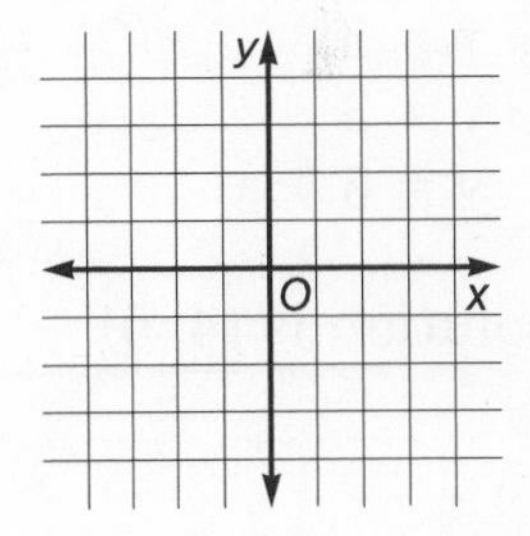

4. $y = x$
$y = -2$

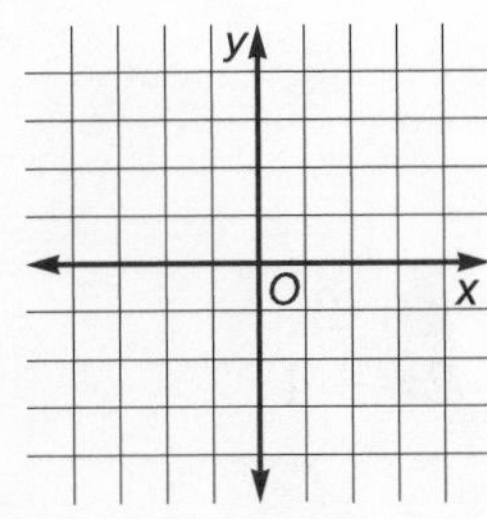

5. $y = -x - 1$
$y = -3x - 3$

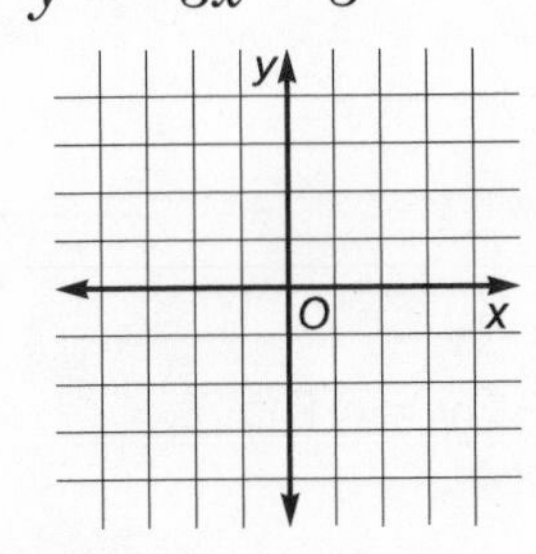

6. $y = 6x - 12$
$y = 2x - 4$

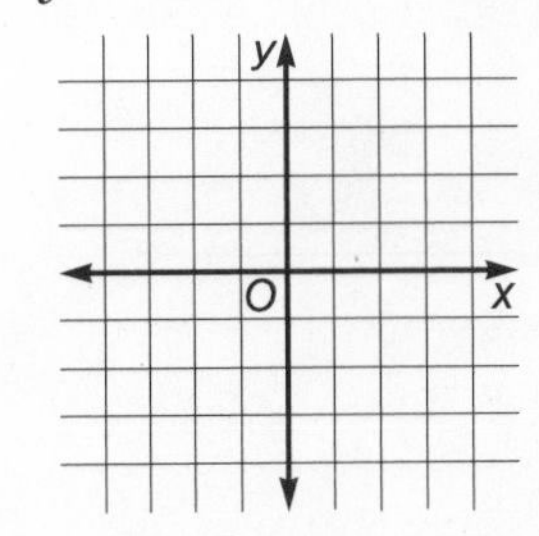

16-2

NAME ______________________ DATE __________ PERIOD ______

Study Guide

Student Edition
Pages 681–686

Solving Systems of Equations by Using Algebra

Two linear equations form a **system of equations**. Two frequently used algebraic methods for solving systems of equations are the **substitution method** and the **elimination method**.

Example: Use substitution and elimination to solve the system of equations.

$$2x - y = 3$$
$$3x + 2y = 22$$

Substitution Method

Solve the first equation for y in terms of x.

$2x - y = 3$
$y = 2x - 3$

Substitute $2x - 3$ for y in the second equation and solve for x.

$3x + 2(2x - 3) = 22$
$3x + 4x - 6 = 22$
$7x = 28$
$x = 4$

Find y by substituting 4 for x in the first equation.

$2(4) - y = 3$
$8 - y = 3$
$y = 5$

The solution is (4, 5).

Elimination Method

$2x - y = 3$ *Multiply by 2.*
$3x + 2y = 22$

$4x - 2y = 6$
$3x + 2y = 22$ *Add to eliminate y.*
$7x = 28$
$x = 4$

Substitute 4 for x in the first equation. Solve for y.

$2(4) - y = 3$
$y = 5$

State whether* substitution *or* elimination *would be better to solve each system of equations. Then solve the system.

1. $y = 3x - 1$
$8 = 2x + 3y$

2. $2x + y = 13$
$5x + y = 28$

3. $3x + y = 10$
$4x - y = 4$

4. $y = x + 1$
$x + y = 5$

5. $y = 2x$
$x - 3y = 10$

6. $y = x + 2$
$y = 2x - 1$

NAME ______________________ DATE ________ PERIOD _____

Study Guide

Student Edition
Pages 687–691

Translations

To **translate** a figure in the direction described by an ordered pair, add the ordered pair to the coordinates of each vertex of the figure.

Example: Graph $\triangle ABC$ with vertices $A(-2, 2)$, $B(-1, -2)$, and $C(-6, 1)$. Then find the coordinates of its vertices if it is translated by $(7, 3)$. Graph the translation image.

$A(-2, 2) + (7, 3) \rightarrow A'(5, 5)$
$B(-1, -2) + (7, 3) \rightarrow B'(6, 1)$
$C(-6, 1) + (7, 3) \rightarrow C'(1, 4)$.

The vertices of the translated figure are $A'(5, 5)$, $B'(6, 1)$, and $C'(1, 4)$.

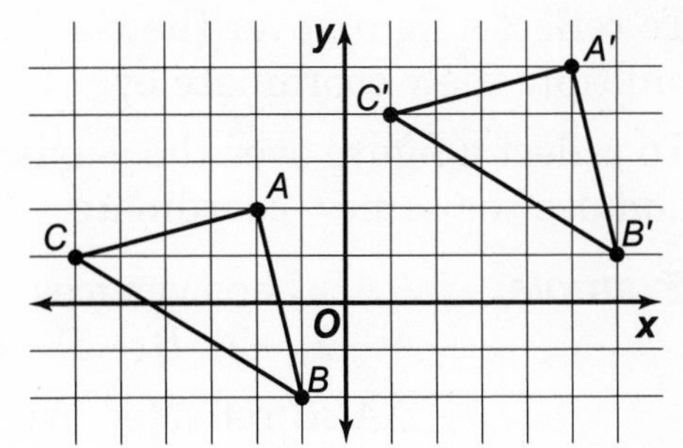

Find the coordinates of the vertices of each figure after the given translation. Then graph the figure and its translation image.

1. $\triangle XYZ$ with vertices $X(-1, 2)$, $Y(2, 3)$, and $Z(3, -1)$, translated by $(-2, -3)$

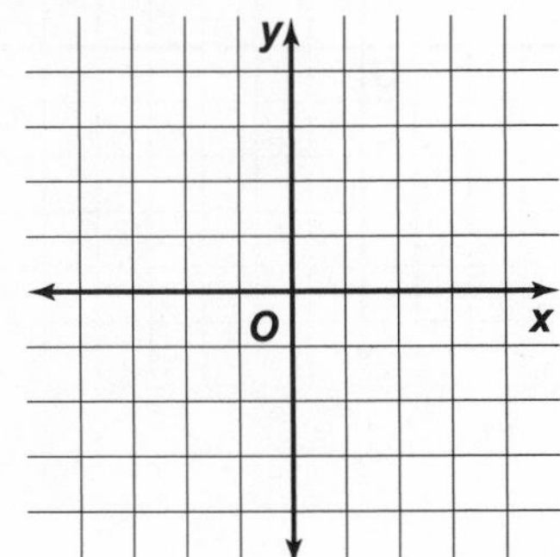

2. polygon $KLMN$ with vertices $K(-1, 1)$, $L(-3, 0)$, $M(-2, -3)$, $N(0, -2)$, translated by $(4, 3)$

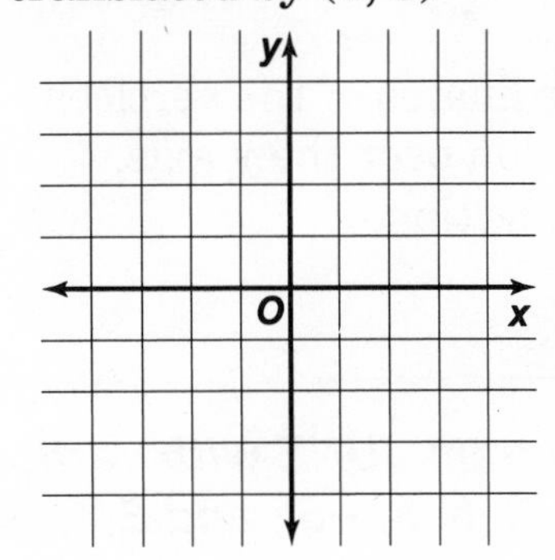

Find the coordinates of the vertices of each figure after the translation described.

3. $\triangle DEF$ with vertices $D(0, 5)$, $E(-1, 3)$, and $F(-3, 4)$, translated by $(2, -1)$

4. pentagon $ABCDE$ with vertices $A(4, -1)$, $B(3, 2)$, $C(1, 4)$, $D(-2, 1)$, and $E(-3, -3)$, translated by $(-2, 1)$

NAME ______________________________ DATE __________ PERIOD ______

Study Guide

Student Edition
Pages 692–696

Reflections

When a figure is **reflected** on a coordinate plane, every point of the figure has a corresponding point on the other side of the line of symmetry.

To reflect a figure over the x-axis, use the same x-coordinate and multiply the y-coordinate by -1.

To reflect a figure over the y-axis, multiply the x-coordinate by -1 and use the same y-coordinate.

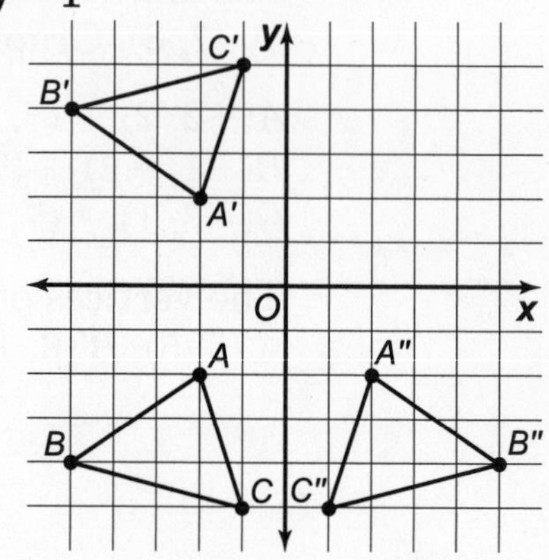

Example: $\triangle ABC$ has vertices $A(-2, -2)$, $B(-5, -4)$, $C(-1, -5)$.

$\triangle ABC$ reflected over the x-axis has vertices $A'(-2, 2)$, $B'(-5, 4)$, $C'(-1, 5)$.

$\triangle ABC$ reflected over the y-axis has vertices $A''(2, -2)$, $B''(5, -4)$, $C''(1, -5)$.

Graph trapezoid BIRD with vertices B(1, 1), I(2, 4), R(6, 4), and D(7, 1).

1. Find the coordinates of the vertices after a reflection over the x-axis. Graph the reflection.

2. Find the coordinates of the vertices after a reflection over the y-axis. Graph the reflection.

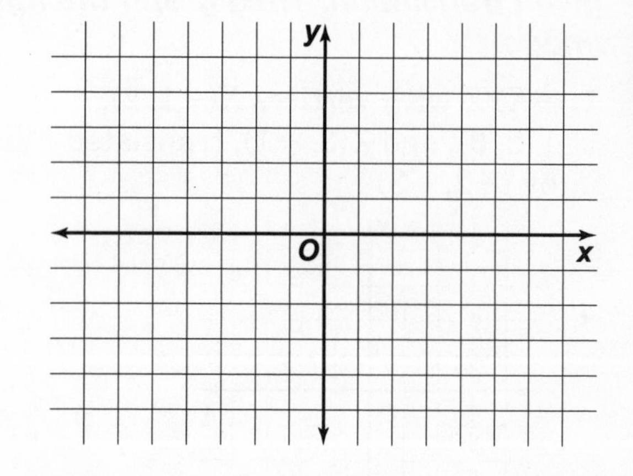

Graph parallelogram JUNE with vertices J(2, −2), U(6, −2), N(8, −5), and E(4, −5).

3. Find the coordinates of the vertices after a reflection over the x-axis. Graph the reflection.

4. Find the coordinates of the vertices after a reflection over the y-axis. Graph the reflection.

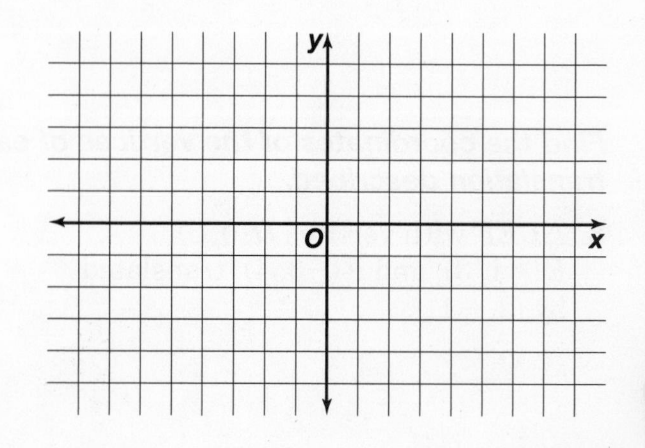

NAME ______________________ DATE ________ PERIOD ______

Study Guide

Student Edition
Pages 697–702

Rotations

Triangle XYZ has vertices $X(-4, 1)$, $Y(-1, 5)$, and $Z(-6, 9)$.

To **rotate** $\triangle XYZ$ 180°, multiply each coordinate by -1.

$X(-4, 1) \rightarrow X'(4, -1)$
$Y(-1, 5) \rightarrow Y'(1, -5)$
$Z(-6, 9) \rightarrow Z'(6, -9)$

To rotate $\triangle XYZ$ 90° counterclockwise, switch the coordinates and multiply the first by -1.

$X(-4, 1) \rightarrow X''(-1, -4)$
$Y(-1, 5) \rightarrow Y''(-5, -1)$
$Z(-6, 9) \rightarrow Z''(-9, -6)$

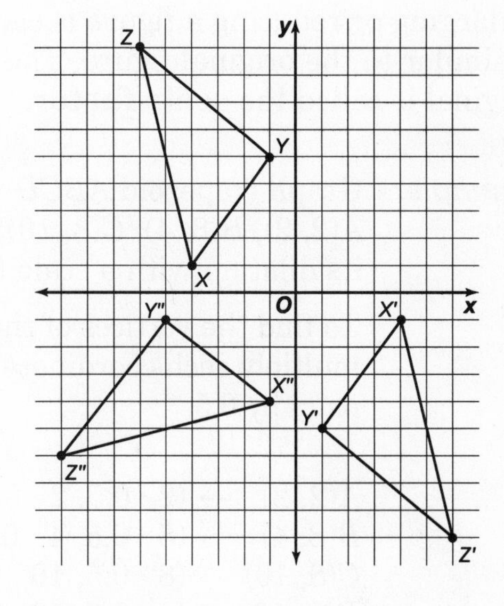

Triangle RST has vertices R(−2, −1), S(0, −4), and T(−4, −7).

1. Graph $\triangle RST$.
2. Find the coordinates of the vertices after a 90° counterclockwise rotation. Graph the rotation.
3. Find the coordinates of the vertices after a 180° rotation. Graph the rotation.

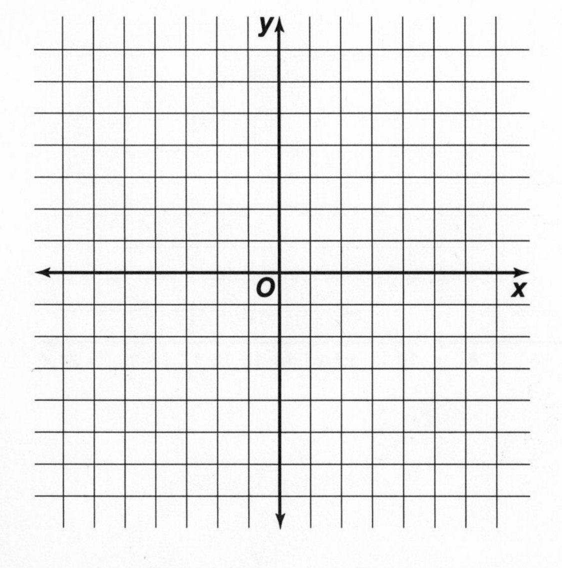

Rectangle TWIN has vertices T(2, 1), W(6, 3), I(5, 5), and N(1, 3).

4. Graph rectangle $TWIN$.
5. Find the coordinates of the vertices after a 90° counterclockwise rotation. Graph the rotation.
6. Find the coordinates of the vertices after a 180° rotation. Graph the rotation.

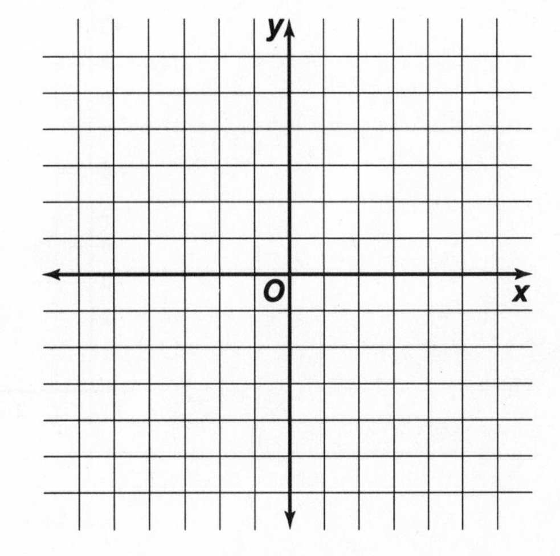

NAME ______________________ DATE ________ PERIOD _____

Study Guide

Student Edition
Pages 703–707

Dilations

Enlarging or reducing a figure is called a **dilation**. A dilated figure is similar to the original figure. The ratio of the new figure to the original is called the **scale factor**.

Example: Graph trapezoid $ABCD$ with vertices $A(2, 2)$, $B(8, 4)$, $C(8, 10)$, $D(2, 10)$. Graph its dilation with a scale factor of 0.5.

To find the vertices of the dilation image, multiply each coordinate in the ordered pairs by 0.5.

$A(2, 2) \rightarrow (2 \cdot 0.5, 2 \cdot 0.5) \rightarrow A'(1, 1)$
$B(8, 4) \rightarrow (8 \cdot 0.5, 4 \cdot 0.5) \rightarrow B'(4, 2)$
$C(8, 10) \rightarrow (8 \cdot 0.5, 10 \cdot 0.5) \rightarrow C'(4, 5)$
$D(2, 10) \rightarrow (2 \cdot 0.5, 10 \cdot 0.5) \rightarrow D'(1, 5)$

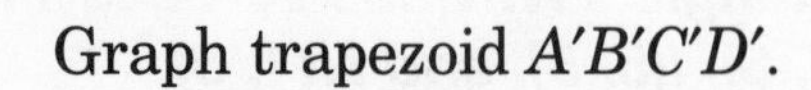
Graph trapezoid $A'B'C'D'$.

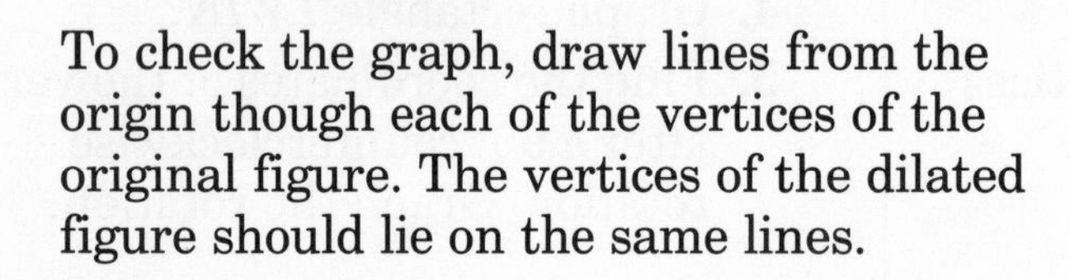
To check the graph, draw lines from the origin though each of the vertices of the original figure. The vertices of the dilated figure should lie on the same lines.

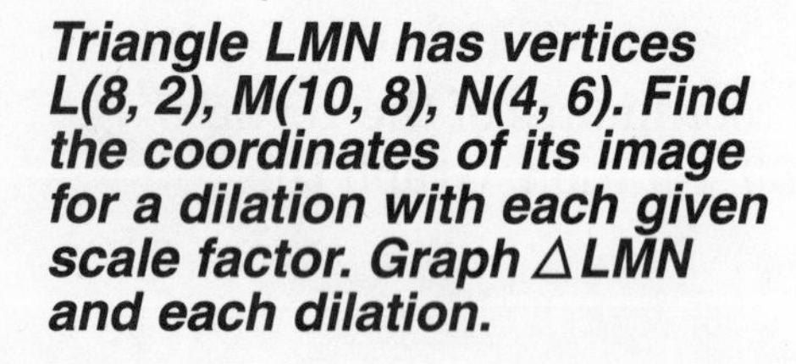
Triangle LMN has vertices L(8, 2), M(10, 8), N(4, 6). Find the coordinates of its image for a dilation with each given scale factor. Graph △LMN and each dilation.

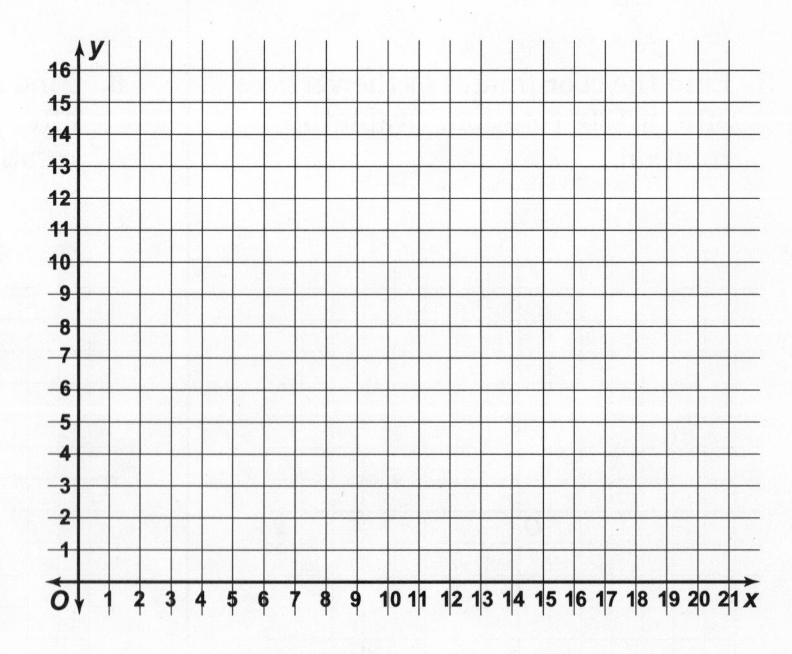

1. 0.5

2. 1.5

3. 2